HIGHER GEOGRAPHY
Applications

Editor: **Martin Duddin**

Morven Archer Jim Bruce
Gordon Lobban and **David Russell**

Hodder & Stoughton
A MEMBER OF THE HODDER HEADLINE GROUP

Acknowledgements

The authors would like to thank the following individuals and organisations for help given during the writing of this book: Angus Murray, Jim Loring and Nick Burn from the Tear Fund; Edwin J Best (Jnr) Librarian, Tennessee Valley Authority; Kerry Watt, Librarian, University of Edinburgh; Eunice Page, Headteacher, West Burton Primary School; Stan Whitaker, Scottish Natural Heritage; Bill Wood, Education Officer, Yorkshire Dales National Park Authority; Dave Parrish, Minerals Planning Officer, Yorkshire Dales National Park Authority; Jim Ferguson, Area Planning Officer, Yorkshire Dales National Park Authority; Robert Moore, Estates Manager, Tilcon Limited; Adekunle Adeoye, Rural and Marine Environment Division, MAFF; South Downs Conservation Board; Countryside Agency; City of Toronto Planning Authority; Alan Doherty, Malcolm Macdonald, Dave Orem, Laura Girdwood and Mabel Black.

The publishers would like to thank the following for permission to reproduce photographs in this book. Every effort has been made to trace and acknowledge ownership of copyright. The publishers will be glad to make suitable arrangements with any copyright holders whom it has not been possible to contact.

Corbis Ecoscene (1.1)/ Robert Estall (1.2)/ Eric Crichton (1.5)/ John Heseltine (1.8); Robert Harding Picture Library (1.10); Corbis/Cordaiy Photo Library (1.17)/ Robert Estall (1.18); Kenneth Maclean (1.19); Countryside Agency (1.20); Imperial War museum (1.22); Brighton & Hove Argos (1.25, 1.26); Robert Harding Picture Library/Ian Grifffiths (1.27); Corbis/Michael John Kielty (1.33)/ Patrick Ward (1.34)/ Adam Woolfitt (1.35)/ Robert Pickett (1.36a, 1.36b); Tony Waltham Geophotos Nottingham (1.39); Hanson aggregates (1.42); Leslie Garland Picture Library/Andrew Lambert (1.50)/ Colin Raw (1.51); Corbis/Andrew Brown (1.58); Still Pictures/Dylan Garcia (1.60); Davnik Photography/David Gowans (1.62); Stock Scotland/Steve Austin (1.66); Scottish Natural Heritage/Peter Barr (1.67); Corbis/Wildcountry (1.70a, 1.70c)/ Roger Antrobus (1.70b)/ Robert Pickett (1.70d); BBC (1.75); Still Pictures/Dylan Garcia (1.77); Scottish Natural Heritage/Lorna Gill (1.78); RSPB (1.79); Corbis (2.8)/ National Archives (2.20)/ Jim Richardson (2.24)/ Bettmann (2.35 both); Franklyn D. Roosevelt Library (2.25, 2.27); Corbis/Library of Congress (2.26); Holt Studios (2.38, 2.39); Still Pictures/Jim Wark (2.41)/ Mark Edwards (2.47, 2.50, 2.52)/ J.P. Delobelle (2.53); Images Colour Library (3.8); Corbis/Wolfgang Kaehler (3.20); Alan Doherty (3.33, 3.39, 3.40, 3.41, 3.45, 3.46, 3.47; Corbis/Gallo Images (4.2)/ Larry Lee Photography (4.10)/ Macduff Everton (4.13)/ Michael S. Yamashita (4.14); Still Pictures/Ron Giling (4.16)/ Jorgen Schytte (4.17)/ Adrian Arbib (4.21)/ Wolfgang Kaehler (4.22)/ Yann Arthus-Bertrand (4.23); Robert Harding Picture Library (4.24); Still Pictures/Hartmut Schwarzbach (4.25 both); World Health Organisation (4.29); Corbis/Stephanie Colasanti (4.32)/ Lloyd Cluff (4.33); The Tear Fund (4.37–4.41).

All other photos supplied by the author.

Orders: please contact Bookpoint Ltd, 130 Milton Park, Abingdon, Oxon OX14 4SB. Telephone: (44) 01235 827720, Fax: (44) 01235 400454. Lines are open from 9.00 – 6.00, Monday to Saturday, with a 24 hour message answering service. Email address: orders@bookpoint.co.uk

British Library Cataloguing in Publication Data
A catalogue record for this title is available from The British Library

ISBN 0 340 78267 6

Published by Hodder & Stoughton Educational Scotland
First published 2001

Impression number	10 9 8 7 6 5 4 3 2
Year	2006 2005 2004 2003 2002

Cover photos from Photodisc, Corbis and Hanson Aggregates.
Typeset by Dorchester Typesetting Group Ltd, Dorchester, Dorset
Printed in Great Britain for Hodder & Stoughton Educational, a division of Hodder Headline Plc, 338 Euston Road, London NW1 3BH by The Bath Press Ltd.

Contents

Editor's foreword

This book has been written to cover the requirements of four of the optional six topics for the Applications of Geography Unit 2 of the National Qualifications Geography Higher syllabus for the Scottish Qualifications Authority. It is further designed to complement the textbook for *Core Higher Geography*, published by Hodder & Stoughton.

The book is designed to replace the successful series of six textbooks *Aspects of Applied Geography* published by Hodder & Stoughton in the mid-1990s. Since most students will study no more than three applications, it was decided to select only the four most popular applications for this textbook. Although in part based on the *Aspects of Applied Geography* texts, the opportunity has been taken to rewrite large sections of the books and to include newer and relevant case study material essential to further developing the key ideas and concepts introduced through the core.

The in-depth case studies, from a variety of Scottish, UK and world locations, as specified in the syllabus, aim to provide students with a variety of resources to allow them to develop skills aimed at:

- making use of relevant geographical terminology, concepts and ideas
- extracting, interpreting, analysing and presenting geographical evidence
- recording information from a variety of sources
- developing communication skills in written, map and graphical forms
- providing geographical evidence and knowledge in order to encourage independent thinking and the ability to make informed judgments.

Each of the four applications is introduced using clear links to the core content of the Arrangements and should allow seamless integration with core units where this is the preferred teaching method. Text-based questions test students' understanding of key ideas and concepts, and key ideas from the syllabus are summarised at the end of each application. A final section gives tips and hints for preparing for the examination and includes examples of exam style questions.

Martin Duddin
June 2001

Rural Land Resources

Links to the core

This section builds on the ideas and concepts introduced as part of the lithosphere section of the physical core unit. It focuses particularly on the formation and characteristic features of three landscape types within the UK: scarp and vale, upland limestone and glaciated uplands. Links to the rural geography unit of the human core are also developed through case studies of rural landscapes and the pressure for change within the rural resource base of the UK.

Rural landscapes provide a variety of physical, economic and social resources. These resources are the product of the interaction between a wide range of physical factors which have been modified by human activity.

Physical factors

The varied rural landscapes of the UK range from the Highlands of Scotland to the East Anglian lowlands (Figures 1.1 and 1.2) and are the result of underlying geology and processes acting upon it (Figure 1.3). Although the UK covers a relatively small area, rock types change considerably over short distances and this results in a variety of distinctive landscapes.

As Figure 1.3 indicates, the older and more resistant granitic, **igneous** and **metamorphic rocks** tend to be found in the north and west of the UK.

Figure 1.1 Scottish Highlands. **C** (This icon indicates that this figure can be found in the colour section on page 189.)

Figure 1.2 East Anglia. C

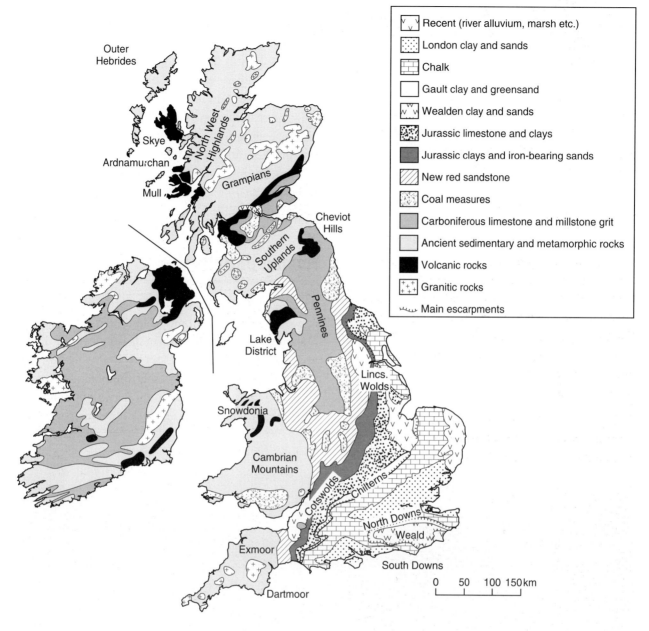

Figure 1.3   The geology of the British Isles.

These are familiar to us as the rocky coastlines and moorlands of south west England, and the mountain massifs of Wales, the English Lake District and Scotland. Elsewhere, particularly in eastern and southern UK, the landscape is based on **sedimentary rocks**, an accumulation of mud, shales, sand and shells formed from pre-existing rocks. These rocks include the upland limestones of the Pennines and the chalk downs of southern England.

It is important to remember that any landscape is not static, but is subject to constant change. Although the

UK is not currently subject to dynamic elements of landscape change, such as volcanic eruptions and active glaciers, these have been significant in the formation of the landscape in the past. Long term processes of landscape change are still at work today, and these are no less significant in altering the land surface. These changes occur so slowly that it is seldom possible to monitor major changes within a human lifetime. The shape of the scenery which we see around us is the result of surface processes; the **weathering** and **erosion** of rock, and its subsequent **transportation** and **deposition** (Figure 1.4).

Processes of landscape change

Weathering
All processes - mechanical, chemical and biological - by which rocks are loosened and disintegrated by exposure to the elements

Mass Movements
Downward movement of weathered material on a slope under the influence of gravity and usually lubricated by rainwater or snowmelt

Erosion
The sculpting of the Earth's surface and scenery by water, wind or ice movement

Transportation
Removal and movement of rock debris by the agents of erosion

Deposition
The laying down of material transported by running water, wind, ice and the sea

Figure 1.4   The physical processes of landscape change.

Although glaciation may have been the most important single factor in shaping the landscape of the Cairngorm mountains in Scotland, surface processes are still at work today modifying the mountain scenery. At the other end of the UK, in the chalk downs of Sussex, the combined effects of rock type and structure, and surface processes can be seen. Here the erosion of a dome of rock uplifted during the Alpine mountain building period, and its subsequent erosion has created the distinctive scarp and vale scenery of alternating steep and gentle chalk slopes.

Human activity

Although the geology and surface processes described above have determined the main characteristics of our rural landscape, the human development of the economic potential of the land has been equally important in shaping the present appearance of the rural landscape. Of the economic activities found in rural areas, farming, both arable and pastoral, has had

Figure 1.5   Rural Britain.

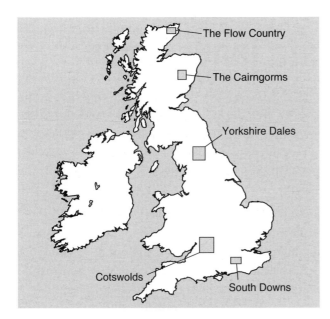

Figure 1.6   Selected rural landscapes of the UK.

the most significant influence on landscape development, and agriculture remains the predominant land use in rural areas today.

Since the 1851 census, the majority of the UK population have been urban dwellers living in towns and cities. Ever since this rural–urban population shift took place, '. . . popular perceptions of rural life have had little in common with reality, focusing instead on idealised cultural images of continuity, stability and changelessness' (The National Trust) (Figure 1.5). This nostalgic image ignores the reality that rural landscapes are living entities, peopled by individuals and communities. The countryside is a dynamic place, facing constant change and competing demands on its resources from pressures as diverse as agriculture, quarrying, forestry, water supply, tourism and recreation. Sometimes such change is slow and gradual, i.e. the decline in village services and changing employment structures, but sometimes it is much faster, i.e. the expansion of forestry in the Scottish Flow Country during the 1980s.

Every landscape has the potential for change. Some changes may be positive and desirable, others potentially damaging, others still may be both desirable and damaging and result in major conflicts between different interest groups. All of these issues are examined in the following sections of this chapter.

The resultant rural landscapes

The sections which follow look in detail at the evolution, development and current pressures on a number of very distinct rural landscapes types in different parts of the UK (Figure 1.6):

- The scarp and vale scenery of southern England, focusing on the Cotswolds area of Jurassic limestone and the chalklands of the Sussex Downs.
- The carboniferous limestone landscapes of the northern Pennines, with case study material drawn from the Yorkshire Dales National Park.
- The landscapes resulting from glaciation in the Scottish Highlands, contrasting the landscape and land use of the Cairngorm mountains with the Flow Country of Caithness and Sutherland.

Questions

1 Using Figures 1.1, 1.2 and 1.3, describe and account for the differences in landscape and relief between the north west and south east of the UK.

2 Identify and describe:
   a the main physical processes,
   b the main human processes of landscape change in Britain today.

3 How does the nostalgic image of rural Britain ignore the reality?

1.2 Scarp and vale landscapes

Jurassic limestone and **chalk** are both **permeable rocks** which allow water to pass through them and which occur as upland ridges throughout parts of England. The uplands usually consist of sharply defined **escarpments** or **scarps** with steep edges, and a more gentle **dip slope** on the other side. The slopes often give way to valleys made up of **impermeable** clays, sometimes occupied by rivers. Together these uplands and valleys give rise to distinctive scarp and vale landscapes.

Jurassic limestone: the Cotswolds

Jurassic limestones form hills which extend from the North Sea coast (between Redcar and Filey in North Yorkshire), through the Midlands to the Dorset coast

at Lyme Bay (Figure 1.7). In the north east, the rocks have only a moderate dip slope and the resulting landforms are **dissected plateaus.** In the Cotswolds they form a system of distinctive escarpments and valleys.

Jurassic limestone was deposited on the bed of a shallow sea which covered much of the UK during the Jurassic geological period. Limestone is harder than chalk, and forms rock outcrops with **bedding planes** and **joints** or cracks which allow water to pass through. The rock is **oolitic**, consisting mainly of minute particles of calcium carbonate mixed with various types of shells, including corals which were formed during a period of sub-tropical climate. The rock colour varies from silver-white to yellow-brown (depending on the iron oxide content), and is much prized as a building stone. Although found in several parts of England, Jurassic limestone has perhaps its clearest identity in the Cotswolds. The importance of this rock type can be gauged from this assessment of landscape quality carried out during the 1990s:

'*Nowhere else in Britain has the underlying rock had such a dominant and unifying effect on the landscape and architecture as in the Cotswolds. As a result, topography, vegetation and settlement are uniquely blended and in harmony. This unity is emphasised by the recurrent visual themes of stone walls, sheep walks and wolds; ancient beech woods; hill forts and Roman roads; woollen towns and cottage gardens; manor houses and parks; water mills and water meadows. It is reinforced by the visually dominant scarp and the Jurassic Ridgeway that stretch the full length of the area. Most of all, though, it is seen in the fact that entire villages and whole small towns are constructed of nothing but oolitic limestone, with not a brick or slate in sight.*' (Figure 1.8)

SOURCE: 'THE COTSWOLD LANDSCAPE'
COUNTRYSIDE COMMISSION (1990)

Figure 1.7 Scarp and vale scenery in England.

Figure 1.8 General view of The Cotswolds. C

Landscape character

Ascending gradually from the low **clay vales** on the Oxford side to the east, the most dramatic scenery is found on the north west facing slopes which form prominent escarpments overlooking the wide plains of the Vale of Severn (Figure 1.9). Here, the geology of the Cotswolds has had a significant impact on the landscape. As mentioned above, Jurassic limestone is a hard rock and is resistant to erosion, forming escarpments reaching more than 300m. The escarpments have progressively eroded and retreated to the south east following lines of weakness in the rock. This has created deep, wide valleys particularly around Bath and Stroud. In places, harder outcrops of rock have resisted erosion and remain as outlying hills, or **outliers** such as Bredon Hill and Oxenton Hill (Figure 1.10). In the north the scarp face is less pronounced, extending as a long finger of upland to Edge Hill in Warwickshire. On the eastern side, the dip slope is broken up by shallow valleys and rivers such as the Evenlode, Windrush, Leach, Coln and Churn, all of which are tributaries of the River Thames (Figure 1.9).

The drainage pattern of the Cotswolds is closely linked to the geology.

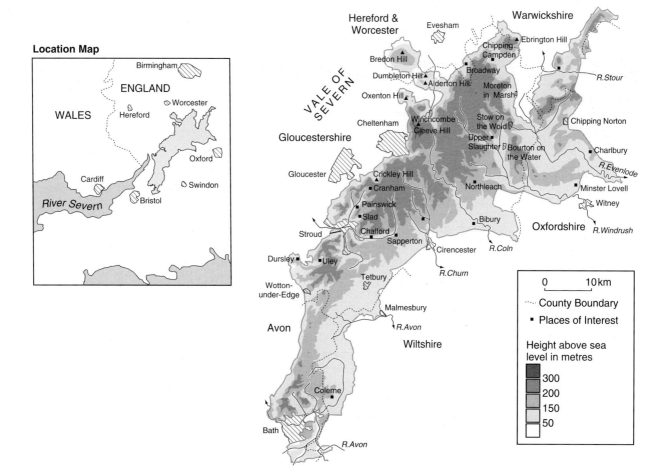

Figure 1.9 Location of the Cotswolds.

Figure 1.10   Cotswolds Outliers – Bredon Hill.

The **differential permeability** of the Jurassic limestone and the underlying **Lias** rocks gives rise to the creation of springs at the junction of these two rock types, most notably on the scarp slope. Villages are often sited along these **spring lines**. On the dip slope, as streams flow over the permeable limestone their length and volume are reduced due to underground seepage, and many stretches of streams only flow during the winter when the **water table** is high, these are called **winterbournes.**

Cotswold soils are dry and alkaline, formed directly from bedrock. There is little high quality agricultural land, but sheep grazing and arable farming are common, with cattle grazing on the less well drained soils in the valleys.

Despite their apparent physical and geological unity, the Cotswolds display a surprising diversity of scenery within a small area (Figure 1.11). The escarpment is linear and dominated by woodland, open common and hill forts. The high wold is characterised by rolling open plateaus contrasting with the more secluded valleys and villages. The dip slope has wide, open valleys, water meadows and gently rolling land.

Landscape change today

In 1966 the Cotswolds were designated an **Area of Outstanding Natural Beauty (AONB)**, in order to conserve the natural beauty of the landscape and control development in the area. The region is immensely popular as a destination for tourists and for the large, highly mobile urban populations which

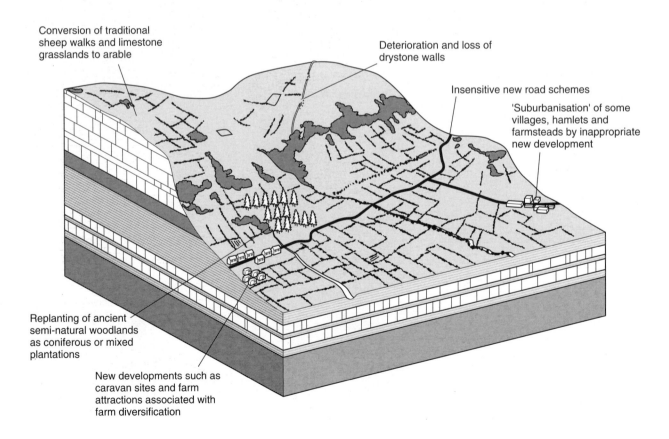

Conversion of traditional sheep walks and limestone grasslands to arable

Deterioration and loss of drystone walls

Insensitive new road schemes

'Suburbanisation' of some villages, hamlets and farmsteads by inappropriate new development

Replanting of ancient semi-natural woodlands as coniferous or mixed plantations

New developments such as caravan sites and farm attractions associated with farm diversification

Figure 1.12   The Cotswolds: vulnerability to change. Ⓒ

are in close proximity. Although there are few immediate landscape threats, it is possible to identify some emerging trends which, if allowed to develop further, could threaten the integrity of the Cotswold landscape for future generations. These pressures are summarised in Figure 1.12.

Designation of Environmentally Sensitive Areas (ESAs)

Incentives to increase food production in the past have led to changes in farming practices in the UK. There has been a shift from mixed farming and traditional stock rearing to more intensive methods of livestock farming and crop production.

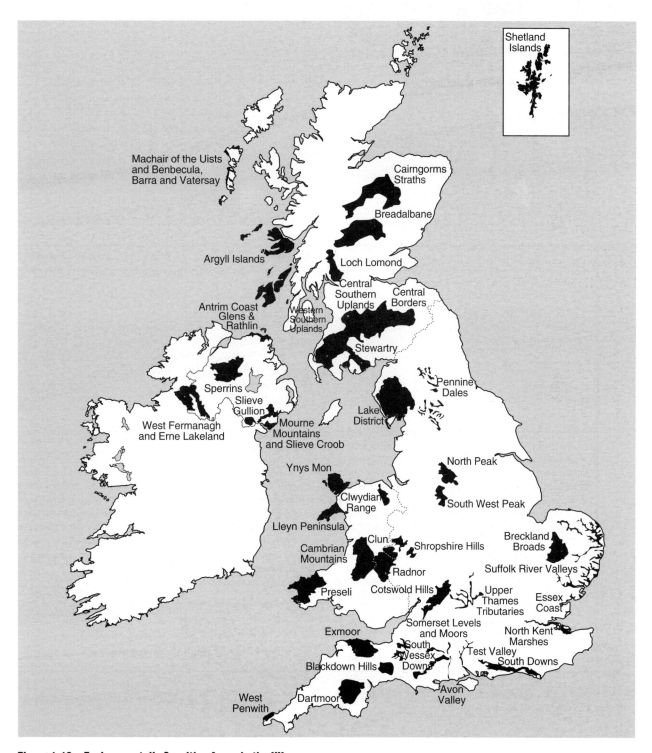

Figure 1.13   Environmentally Sensitive Areas in the UK.

Key

■ Landslip		▨ Upper Lias	
▨ Great Oolite	Mid Jurassic	▨ Middle Lias	Lower Jurassic
▨ Inferior Oolite		▨ Lower Lias	

0   1 km

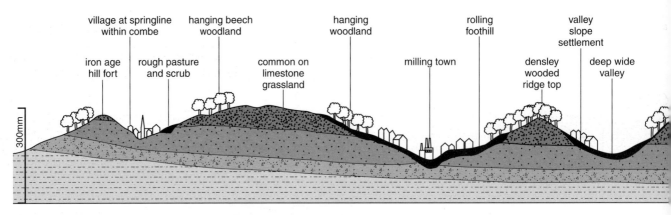

village at springline within combe   hanging beech woodland   hanging woodland   rolling foothill   valley slope settlement

iron age hill fort   rough pasture and scrub   common on limestone grassland   milling town   densley wooded ridge top   deep wide valley

300mm

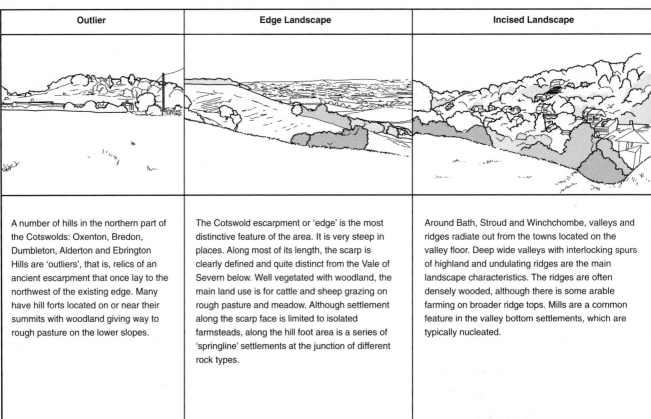

Outlier	**Edge Landscape**	**Incised Landscape**
A number of hills in the northern part of the Cotswolds: Oxenton, Bredon, Dumbleton, Alderton and Ebrington Hills are 'outliers', that is, relics of an ancient escarpment that once lay to the northwest of the existing edge. Many have hill forts located on or near their summits with woodland giving way to rough pasture on the lower slopes.	The Cotswold escarpment or 'edge' is the most distinctive feature of the area. It is very steep in places. Along most of its length, the scarp is clearly defined and quite distinct from the Vale of Severn below. Well vegetated with woodland, the main land use is for cattle and sheep grazing on rough pasture and meadow. Although settlement along the scarp face is limited to isolated farmsteads, along the hill foot area is a series of 'springline' settlements at the junction of different rock types.	Around Bath, Stroud and Winchchombe, valleys and ridges radiate out from the towns located on the valley floor. Deep wide valleys with interlocking spurs of highland and undulating ridges are the main landscape characteristics. The ridges are often densely wooded, although there is some arable farming on broader ridge tops. Mills are a common feature in the valley bottom settlements, which are typically nucleated.

Figure 1.11   The landscape character of the Cotswolds.

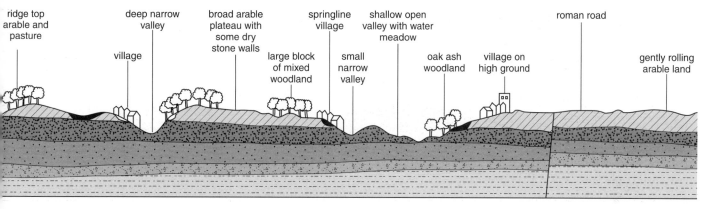

ridge top arable and pasture

deep narrow valley

broad arable plateau with some dry stone walls

springline village

shallow open valley with water meadow

roman road

village

large block of mixed woodland

small narrow valley

oak ash woodland

village on high ground

gently rolling arable land

Valleys and Ridges	High Wold and Wold	Dip Slope
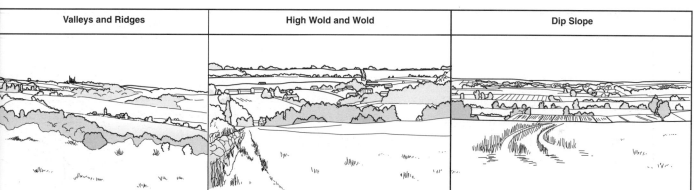		
Mid-way between the edge and dip slopes, this area is characterised by narrow deep valleys with broad flat intervening ridges. The valleys form pastureland with settlements located on valley slopes. In contrast, the ridges are more open with expansive, spreading suburbanised settlements. Within this zone there is almost a sense of two very different landscapes in juxtaposition.	Found principally towards the northern part of the Cotswolds, here the topography is softer and valleys narrower with broad expansive plateau tops with large-scale open landscapes. Nucleated springline villages are often centred around features such as greens and commons. Some parts are unspoilt but in certain areas there has been sporadic development such as electricity pylons, radio masts, airfields and unsympathetic farm buildings which have caused visual intrusion into the landscape.	South and east of the Wolds the landscape becomes more gentle and open along the dip slope which covers a large area along the eastern side of the Cotswolds. The plateau is dissected by rivers, such as the Windrush and Coln, which occupy broad valleys. Arable farming is the dominant land use with large fields separated by hedgerows or fences. The landscape is open with few areas of woodland although there are some recent conifer plantations. Settlements may occupy 'dry point' sites above the river floodplains. Due to their ease of access and proximity to larger towns there has been some 'suburbanisation' of dip slope settlements with expansion of modern housing.

In many instances this has led to a loss of wildlife habitats and a loss of traditional landscape features. The rolling landscape of the Cotswold Hills, with its large fields enclosed by traditional drystone walls and hedges is an example of a distinctive farming landscape which has evolved in sympathy with the local limestone environment. In 1994 the Cotswolds were designated an **Environmentally Sensitive Area (ESA)** where farmers are offered financial incentives not only to conserve, but also to enhance, and where possible, recreate valued landscape features such as drystone walls and wildlife habitats. Additionally, farmers are encouraged to provide new opportunities for public access for walking and other quiet recreation.

In the Cotswolds much of the environmental value of the area today is based on traditional farming practices associated with sheep grazing. The aim of the ESA scheme is 'to maintain and enhance the traditional landscape and wildlife interest and to conserve and protect the area's archaeological and historic features' (MAFF). Farmers who enter the scheme must agree to maintain existing grassland and not to increase the area of arable farmland. They must also restrict their use of pesticides and fertilisers, and restore drystone walls using traditional techniques and materials. Additionally, the scheme encourages farmers to revert arable land to extensively managed permanent grassland.

So far, in both the Cotswolds and the other ESAs in the UK (Figure 1.13) the scheme has been very successful. Other European Union (EU) countries have already followed the UK in introducing ESA schemes, as all Member States are required to introduce similar measures under the Agri-Environment Regulation adopted in 1992. This offers financial incentives from the EU budget to encourage farmers to manage their land in ways that benefit the environment.

Questions

1 Using Figure 1.11 describe and explain the main contrasts between the landscapes and land-uses of the Cotswold escarpment and the dip slope.

2 What are the main threats to the Cotswold landscape and how can they be minimised?

Chalk landscapes: the South Downs

Chalk outcrops are found mainly in southern and eastern England forming distinctive scenery known as **downland**. Chalk is a soft white limestone made up almost entirely of calcium carbonate. It originates from the skeletal remains of marine creatures deposited, compressed and eventually fossilised on the sea-bed to form chalk. As well as being permeable it is also a **porous** rock, allowing water to pass through tiny pores within its structure. Areas of chalk downland in the UK are shown in Figure 1.7 and form a series of long upland ridges such as the North and South Downs, or more extensive areas of gently undulating uplands such as Salisbury Plain.

The South Downs (Figure 1.14) are a relic of massive movements in the Earth's crust which took place 20 million years ago. The area which is now south east England was then a vast low plain formed from layers of chalk, clay, silt and sand built up over time and continuously deposited and compressed during the **Cretaceous** geological period. Pressure built up over geological time pushed these layers of rock upwards to form a vast rounded dome, centred on the area now known as the Weald (Figure 1.15).

Gradually, over millions of years, the land changed shape. The centre of the dome was eroded away leaving an outer upstanding rim of chalk surrounding a lowland plain formed from older layers of clay and sandstone. Today the outer rim of the chalk forms the uplands of the North and South Downs and the central plain is the Weald. The layers of chalk are tilted at a sharp angle (Figure 1.16) with steep scarp slopes facing inwards, towards the centre of the original dome and the shallow dip slopes face outwards.

The chalk escarpment dominates the landscape of the South Downs. The simple structure of the rock is revealed in its striking profile where the escarpment reaches the sea at Beachy Head (Figure 1.17). Elsewhere, the chalk forms an area of expansive rolling upland (Figure 1.18) and throughout the downlands surface drainage is more or less absent.

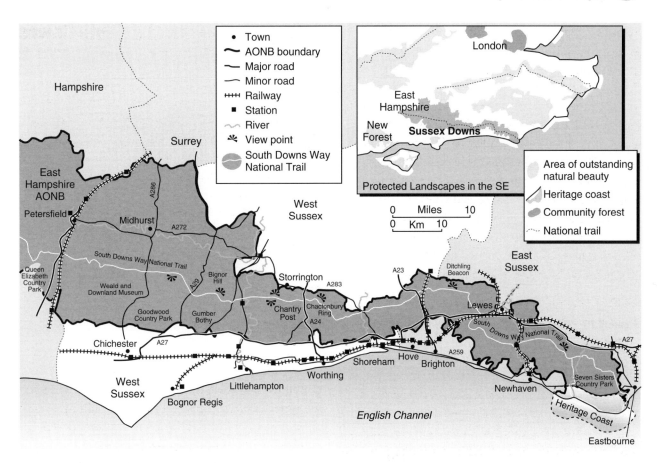

Figure 1.14 **Location of the South Downs.**

The combination of **structural folding, differential erosion** and the **periglacial** conditions experienced by this part of England during the Ice Age has produced distinctive landscape features.

Steep valleys or **coombes** such as the Devil's Dyke (Figure 1.19) cut into the face of the escarpment, and extensive **dry valley** systems are found on the dip slope.

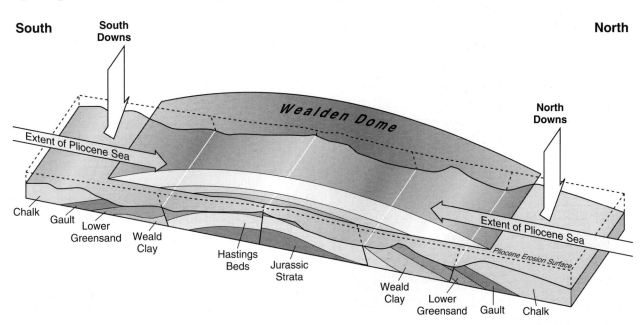

Figure 1.15 **Cross section of the Wealden Dome and escarpments.** Ⓒ

South   **North**

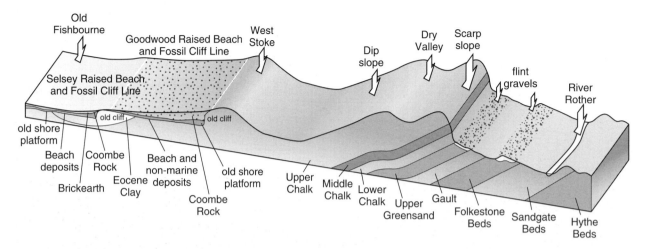

Figure 1.16   The physical structure of the chalklands: escarpments, dip slope and valleys. C

Figure 1.17   Beachy Head.

Geomorphologists believe these valleys were cut into the underlying rock, by streams which flowed when the water table was higher than today.

The South Downs are cut by several major river valleys, the largest of which are occupied by the rivers Arun, Adur, Ouse and Cuckmere (Figure 1.14). These four rivers have created 'gaps' (Figure 1.20) or U-shaped valleys in the escarpment and are valuable routeways through the chalk. Their extensive floodplains, on underlying clay, create the vales which contrast strongly with the dry valleys and expansive arable farmlands of the chalk downlands. The main features of the South Downs landscape are summarised in Figure 1.21.

Forces for change in the landscape

All landscapes are constantly changing, evolving and developing as a consequence of both natural and human-induced processes. Rates of change vary widely. Some, such as the gradual erosion of the Downs themselves by wind and water are slow, whereas the coastal erosion at Beachy Head is very much in evidence with the coast retreating at up to one metre each year. There is no doubt that people have had a far more immediate impact on the environment of the South Downs than any natural process, and there is little reason to believe that this pace of change will decrease in the future.

Figure 1.18   Landscape of the South Downs. Ⓒ

Figure 1.19   Devil's Dyke.

Figure 1.20   A river gap.

The South Downs landscape has always been seen as that of 'real' England and during the Second World War it was even the subject of a propaganda poster (Figure 1.22).

In 1966 the area was designated an Area of Outstanding Natural Beauty (AONB). This was in recognition of the national importance of its scenic beauty, and the particular vulnerability of the downland in the face of development pressures.

In 1999 it was announced that the South Downs would be designated a National Park with a final decision due in 2002. Such additional protection will be welcomed by many but in the meantime, pressures on the South Downs landscape continue in several key areas:

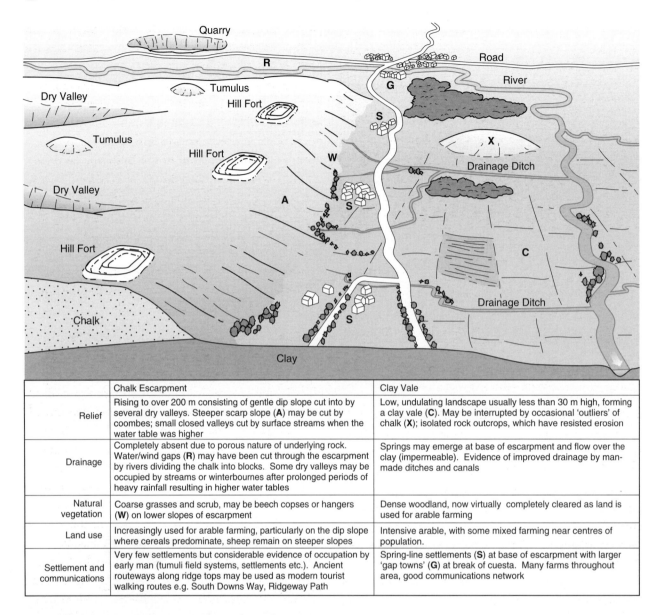

	Chalk Escarpment	Clay Vale
Relief	Rising to over 200 m consisting of gentle dip slope cut into by several dry valleys. Steeper scarp slope (**A**) may be cut by coombes; small closed valleys cut by surface streams when the water table was higher	Low, undulating landscape usually less than 30 m high, forming a clay vale (**C**). May be interrupted by occasional 'outliers' of chalk (**X**); isolated rock outcrops, which have resisted erosion
Drainage	Completely absent due to porous nature of underlying rock. Water/wind gaps (**R**) may have been cut through the escarpment by rivers dividing the chalk into blocks. Some dry valleys may be occupied by streams or winterbournes after prolonged periods of heavy rainfall resulting in higher water tables	Springs may emerge at base of escarpment and flow over the clay (impermeable). Evidence of improved drainage by man-made ditches and canals
Natural vegetation	Coarse grasses and scrub, may be beech copses or hangers (**W**) on lower slopes of escarpment	Dense woodland, now virtually completely cleared as land is used for arable farming
Land use	Increasingly used for arable farming, particularly on the dip slope where cereals predominate, sheep remain on steeper slopes	Intensive arable, with some mixed farming near centres of population.
Settlement and communications	Very few settlements but considerable evidence of occupation by early man (tumuli field systems, settlements etc.). Ancient routeways along ridge tops may be used as modern tourist walking routes e.g. South Downs Way, Ridgeway Path	Spring-line settlements (**S**) at base of escarpment with larger 'gap towns' (**G**) at break of cuesta. Many farms throughout area, good communications network

Figure 1.21   The main features of the South Downs landscape.

Agriculture

Throughout history the chalk hills of the South Downs have been managed to meet human needs. It is thought that around 8000 years ago the Downs were part of a great woodland of lime, elm, oak and hazel. Then came settled farming and the woods were cleared, so that by Roman times much of the downland was in agricultural use. In the Middle Ages, as wool became a valuable commodity, sheep grazing became the main land use with sheep numbers peaking during the seventeenth century. Mixed farming then became more prominent as wool prices fell and arable crop rotation was introduced. Even with heavy ploughing for food during the years of

war in the twentieth century, the distinctive flora and fauna of the downland survived.

The wildlife that withstood centuries of gradual agricultural change has been unable to withstand the effects of intensification of farming during the last 50 years. Sheep grazing has declined as the market crashed. Grasslands were ploughed and poor chalk soils enriched with fertilisers for arable monoculture. Fields were treated with herbicides and pesticides so that very little other than the crop survived. Large-scale mechanisation led to hedgerows being stripped out, and with them their habitats and wildlife. In the face of these radical changes to farming practices, the South Downs ESA was set up in 1987.

Figure 1.22   Poster from World War II.

Over 70% of the downland is now protected by ESA agreements which prohibit the use of fertiliser, pay farmers to revert arable land back to grass and down, and encourage correct grazing management to improve the quality of the chalk grassland and prevent the encroachment of scrub.

ESA agreements usually last for ten years and the scheme in the South Downs was relaunched in 1997. Despite the encouraging increase in the management of chalk grasslands, and the conversion of arable land to pasture, not all influence on local farming has been positive. The EU, through supporting cereal prices created massive 'grain mountains' of overproduction. As a result, since 1992 the EU through its **Common Agricultural Policy (CAP)** has required farmers to '**set-aside**' a proportion of their arable land in return for compensation payments.

Set-aside was introduced as part of a programme for controlling the over-production of cereals within the EU. The UK and the rest of the EU normally produce more cereals than are needed by their own consumers. It is expensive to store and export the surplus to other world markets and the subsidies necessary to allow this are both costly for the taxpayer and disruptive to world trade. Subsidised EU exports can also undermine attempts to develop local agriculture in less economically developed countries, but have been used as emergency aid in times of major famine.

In the UK farmers in areas such as the South Downs can claim support payments and in return must 'set-aside' part of their arable land, taking it out of

production. Under the **Agenda 2000** reforms agreed by the EU, the default obligatory rate is 10%. This has had a particularly significant visual impact on the chalk uplands, with blocks of land being abandoned. Market forces have also had an influence on the landscape and have resulted in a more diverse range of crops, e.g. oilseed rape and linseed now add seasonal splashes of colour to the landscape (as these can be grown on set-aside land). Set-aside rules are regularly reviewed, and one significant environmental improvement which came into effect in January 2000 was that 10 metre wide strips of land next to watercourses and lakes can be used as set-aside land. These strips will provide important environmental benefits such as acting as buffer zones and providing corridor habitats for birds and wildlife.

Changing markets and pressures to produce food at cheaper prices has resulted in many smaller farms on the downlands being amalgamated to create more efficient and economic production units. Pressures to supplement farm income are seen throughout the area in **farm diversification** (Figure 1.23) where the aim is to provide alternative sources of income for the farmer. Diversification can include a number of different approaches (Figure 1.24) although many, such as the development of six new golf courses to the north of Brighton, do pose a threat to the traditional Downs landscape. Others, such as the conversion of traditional farm buildings which have become redundant as a result of changing agricultural practice, may be seen either as positive development in increasing farm income, or as negative if their

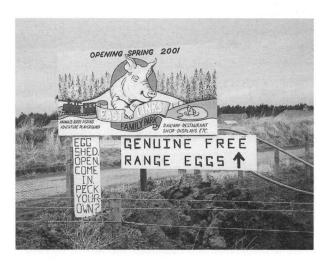

Figure 1.23   Farm Diversification.

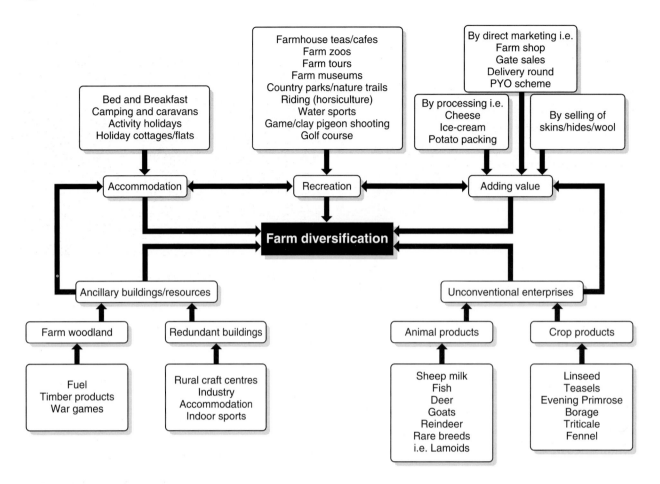

Figure 1.24　Types of farm diversification.

conversion to residential dwellings is accompanied by 'modernisation' out of character with the rural landscape. At the same time the construction of new buildings can have a detrimental visual impact on the wider landscape.

Forestry

Although one of the main attractions of the South Downs landscape is its sense of 'openness', the woodland cover contributes greatly to the landscape character, particularly in the west where it accounts for almost 60% of the land cover. Traditional management techniques which involved thinning and the selective felling of older trees created a diverse woodland mosaic, characterised by patches of woodland of different ages. As local markets for this timber declined in the face of cheaper imports and alternative materials, they were often left unmanaged and became particularly vulnerable to disease and wind damage. Damage caused by the storms of 1987, 1990 and 2000 has exacerbated the problem.

Many derelict woodlands were replanted with conifers after the Second World War, but fortunately their extent is limited and they rarely dominate the landscape over extensive areas, as in other parts of the UK. Since 1991 the **Woodland Grant Scheme** has encouraged the development of broadleaved woodland, and Friston Forest west of Eastbourne has been progressively replanted with broadleaved species. Farm woodlands have also seen improvements since the introduction of the **Farm Woodland Scheme** in 1988. This scheme encourages farmers to plant and maintain farm woodlands. It does this by providing annual payments to farmers who are willing to convert land previously used for crops or grazing, to woodland. Since planting timber is a long-term investment these payments help to support farmers during the growing period. The success of the scheme can be gauged by the rate of planting; over 31 000 hectares have been approved for planting since 1988, some 34 million new trees of which 85% are native broadleaved species such as oak, ash and sycamore. Apart from adding variety to the landscape, the woodlands benefit the economy, environment and farmers as they:

- act as shelter belts for farmland
- offer recreational opportunities
- provide cover for game birds
- over time, produce timber to reduce imports
- provide smaller timber for wood products
- assist in reducing nitrate leaching from farmland.

Realistically, for forestry to be a viable alternative to farming in the South Downs, the financial returns obtained from the trees should be closely related to the potential income from farming. It seems unlikely that the South Downs will see large areas of farmland becoming available for afforestation. The main potential may be in the urban fringe areas which are least attractive to farmers, and on some of the poorest soils.

Mineral extraction

The South Downs have a long history of mineral exploitation. Chalk pits have existed for many centuries reflecting the demand from the farms for lime as a fertiliser. The oldest pits have been well integrated into the local landscape over the years and only where pits have been developed into major quarries, such as at Cocking, south of Midhurst, and around Lewes do they have a major adverse impact on the landscape. Government policy on quarrying within protected landscapes such an Area of Outstanding Natural Beauty (AONB) has become stricter in recent years. Following designation as a National Park, any new application for quarrying in the South Downs will be required to prove that it will have minimal environmental and landscape impact as well as specifying how the site will be restored, even before any quarrying can take place.

Urban development

Classification of the South Downs as an AONB has restricted large scale residential, commercial or industrial development, and designation as a National Park in 2002 will offer increased protection. However, due to the particular geography of the South Downs the visibility of development in the areas adjacent to the chalk escarpment is a significant factor. Pressures for urban development immediately beyond the AONB boundary to the south of the

Figure 1.25   Brighton seen from the South Downs.

Downs have led to a very abrupt interface between the rolling open chalk downlands and the extensive urban conurbations along the coastal plain (Figure 1.25) Associated urban features such as electricity pylons, increased traffic and golf courses seem particularly intrusive in this open, expansive landscape. Most recently, the Brighton by-pass has extended the urban influence further into the Downs, eroding the sense of remoteness.

The scale of modern road improvements and their associated engineering structures mean that such roads have a significant visual impact on the landscape and other significant environmental consequences. The Brighton by-pass was built almost entirely within the AONB, and in places the road has created a scar on the landscape as it slices through ridges and woodlands (Figure 1.26). There are many bridges and junctions which are visually intrusive.

Figure 1.26   Brighton by-pass under construction.

In addition, the increased traffic noise and road lighting has intruded on an area of otherwise quiet countryside and created a new corridor for the focus of further development pressures on the AONB.

There is pressure for minor roads to be upgraded to cope with the increasing volumes of visitors but in fact, the narrow twisting country lanes, high hedge banks and steep gradients are all important local landscape features which are vulnerable to insensitive 'improvements'. Alternative solutions include speed restriction measures such as cattle grids, speed bumps and in some locations, more radical steps such as complete road closures to preserve the sense of remoteness and tranquillity.

Electricity pylons are particularly conspicuous on the open downlands and in recent years the **Countryside Agency**, the government body responsible for conserving the countryside in England, has supported initiatives for underground electricity cables. The cost remains a key issue, although there has been some progress in the downland villages where some intrusive low-voltage lines have been removed.

Recreation and tourism

The South Downs have attracted tourists from the heavily urbanised areas of south east England ever since the opening of the London-Brighton railway in 1841. In addition, the area provides an important recreational resource for the urban areas along the south coast of England and major inland towns such as Guildford, Croydon and Crawley. Recent estimates put the number of visitors to the South Downs each year at around 32 million. Although some parts of the open downland can undoubtedly absorb some of this visitor pressure with minimal disturbance, there is an increasing need for the management of recreation. There are a number of '**honeypot sites**' of intense visitor pressure where the tourists may pose a threat to the very qualities they are seeking - a sense of space and tranquillity. Such sites include historic houses and parks such as Goodwood and Petworth, and certain viewpoints like Devil's Dyke (Figure 1.19). Car parking can be a major problem at these locations particularly on summer weekends, and there are also issues of litter, noise and footpath erosion.

Figure 1.27 Hang Gliding on the South Downs.

Any new National Park authority will need to address these problems urgently and also introduce management strategies for new forms of recreation such as mountain-biking, hang-gliding (Figure 1.27) and four-wheel drive vehicles which may not be compatible with the more traditional, quieter pursuits such as walking and bird watching.

Questions

3 Identify the main similarities and differences between Jurassic limestone and chalk.

4 Using Figure 1.21 describe and explain the main differences in the human use of the scarp slope and clay vale areas.

5 Identify and explain the changes which have taken place in land use on the South Downs since 1950.

6 Explain, with examples, how farmers are diversifying their sources of income, and why such moves have been necessary.

7 In what ways can the development of woodland benefit the South Downs landscape?

8 Give brief details of:
   a Environmentally Sensitive Areas,
   b Set-aside,
   c The Farm Woodland Scheme.
   Explain how they are of benefit to both farmers and conservationists.

9 Identify the main threats to the South Downs landscape and suggest how designation as a National Park will help the area.

1.3 Carboniferous limestone landscapes

Carboniferous limestone landscapes occur throughout the UK (Figure 1.28), from Fife in Scotland to the Mendip Hills in southwest England.

The most extensive outcropping of this distinctive rock type is found in the northern Pennines where the limestone forms upland areas which are separated by flat-floored, and steep-sided glaciated valleys called '**dales**'. The carboniferous limestone found in the Pennines is a hard rock, more resistant to erosion than the Jurassic limestone looked at in Section 1.2. It is grey in colour and occurs in massive blocks separated by joints. These joints allow water to pass through them, making the rock **permeable** and resulting in the surface usually being dry, with streams sinking underground.

The processes of erosion and deposition, and particularly the solution of the limestone by rainwater have produced a distinctive **karst** landscape (Figure 1.29). Although due to the important role of glaciation processes in the development of the present landscape, it is more correctly called **glaciokarst**.

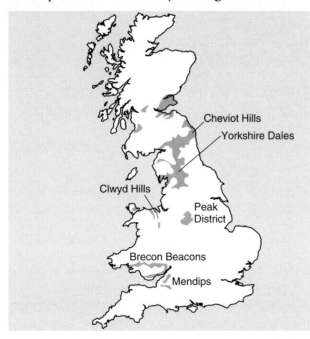

Figure 1.28   The distribution of Carboniferous limestone in the UK.

The Yorkshire Dales National Park

The National Parks of England and Wales were designated by the National Parks and Access to the Countryside Act of 1949. The twin purposes of National Parks were set out in The Environment Act of 1995:

- to conserve and enhance the natural beauty, wildlife and cultural heritage of the areas; and
- to promote opportunities for the understanding and enjoyment of the special qualities of those areas by the public.

There are now 11 National Parks in England and Wales (Figure 1.30) of which the Yorkshire Dales is the third largest, covering an area of 1760 km² (Figure 1.31). The Park has a population of 19 000 people living mainly in small villages. The largest settlement is Sedbergh with a population of 2225. The Park was designated in 1954 primarily to conserve its spectacular Carboniferous limestone scenery, but also the wild moorlands and pastoral valleys (dales) which have given the park its name.

Landscape features and processes

In the area of the Yorkshire Dales National Park, the most impressive features of karst scenery occur in association with the outcrops of **Great Scar Limestone** (Figure 1.32) which was first exposed about two million years ago. This 120–200 m thick slab of rock is almost horizontal and was formed by the gradual deposition of shell debris and chemical precipitates in shallow seas about 330 million years ago during the Carboniferous period. This sediment was compressed over millions of years and recrystallised into limestone which was later cracked and uplifted. The most important process in shaping the limestone scenery has been the repeated climatic changes during the various Ice Ages.

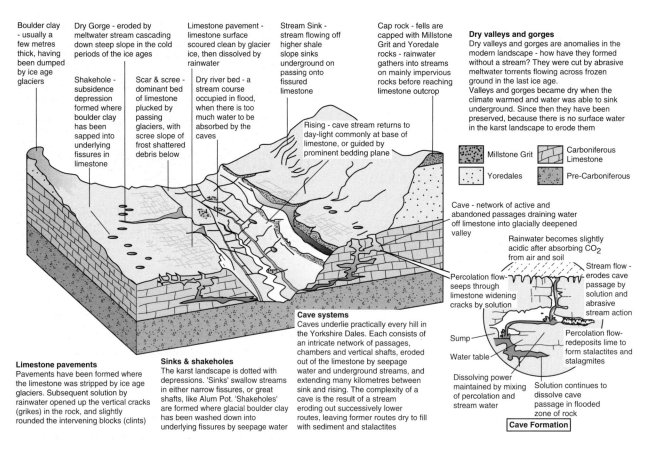

Boulder clay - usually a few metres thick, having been dumped by ice age glaciers

Dry Gorge - eroded by meltwater stream cascading down steep slope in the cold periods of the ice ages

Limestone pavement - limestone surface scoured clean by glacier ice, then dissolved by rainwater

Stream Sink - stream flowing off higher shale slope sinks underground on passing onto fissured limestone

Cap rock - fells are capped with Millstone Grit and Yoredale rocks - rainwater gathers into streams on mainly impervious rocks before reaching limestone outcrop

Dry valleys and gorges
Dry valleys and gorges are anomalies in the modern landscape - how have they formed without a stream? They were cut by abrasive meltwater torrents flowing across frozen ground in the last ice age.
Valleys and gorges became dry when the climate warmed and water was able to sink underground. Since then they have been preserved, because there is no surface water in the karst landscape to erode them

Shakehole - subsidence depression formed where boulder clay has been sapped into underlying fissures in limestone

Scar & scree - dominant bed of limestone plucked by passing glaciers, with scree slope of frost shattered debris below

Dry river bed - a stream course occupied in flood, when there is too much water to be absorbed by the caves

Rising - cave stream returns to day-light commonly at base of limestone, or guided by prominent bedding plane

Millstone Grit

Carboniferous Limestone

Yoredales

Pre-Carboniferous

Cave - network of active and abandoned passages draining water off limestone into glacially deepened valley

Rainwater becomes slightly acidic after absorbing CO_2 from air and soil

Percolation flow - seeps through limestone widening cracks by solution

Stream flow - erodes cave passage by solution and abrasive stream action

Cave systems
Caves underlie practically every hill in the Yorkshire Dales. Each consists of an intricate network of passages, chambers and vertical shafts, eroded out of the limestone by seepage water and underground streams, and extending many kilometres between sink and rising. The complexity of a cave is the result of a stream eroding out successively lower routes, leaving former routes dry to fill with sediment and stalactites

Sump

Water table

Percolation flow - redeposits lime to form stalactites and stalagmites

Dissolving power maintained by mixing of percolation and stream water

Solution continues to dissolve cave passage in flooded zone of rock

Cave Formation

Limestone pavements
Pavements have been formed where the limestone was stripped by ice age glaciers. Subsequent solution by rainwater opened up the vertical cracks (grikes) in the rock, and slightly rounded the intervening blocks (clints)

Sinks & shakeholes
The karst landscape is dotted with depressions. 'Sinks' swallow streams in either narrow fissures, or great shafts, like Alum Pot. 'Shakeholes' are formed where glacial boulder clay has been washed down into underlying fissures by seepage water

Figure 1.29   Karst scenery. C

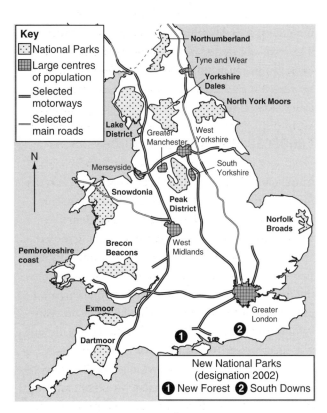

Figure 1.30   National Parks of England and Wales.

Two characteristics of the Great Scar Limestone have been crucial to the development of karst scenery. The first is the rock's chemical weakness, which allows it to be completely dissolved by rainwater.

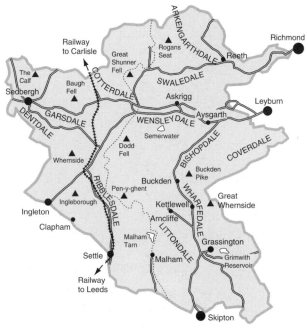

Figure 1.31   The Yorkshire Dales National Park.

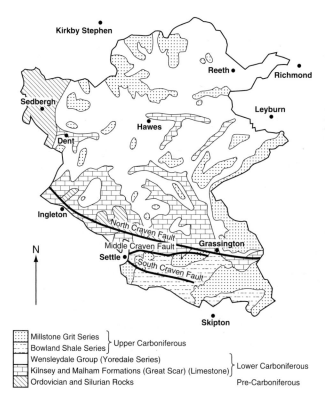

Figure 1.33 Gordale Scar. C

Figure 1.32   Geology of the Yorkshire Dales National Park.

Second, in contrast to the chemical weakness, the rock structure itself is mechanically very strong and this has allowed major cave systems to develop.

The amount of rock dissolved by rainwater depends on the acidity levels of the water percolating down through the rock. This is dependent on carbon dioxide levels absorbed from the surface vegetation. This in turn is largely the product of the climate. During warm periods, when the vegetation is plentiful (like the present day), carbon dioxide levels in the water are high and cave systems are dissolved from the limestone. In contrast, during glacial periods ice became the main agent of erosion, with the deepening of the characteristic glacial valleys. In between these alternating periods of warm and cold climate, **periglacial** conditions prevailed, similar to those found along the edges of the world's tundra regions today. Glacial meltwater is thought to have been responsible for producing some of the most spectacular aspects of limestone scenery, creating **dry valleys** and spectacular **gorges** such as Gordale (Figure 1.33). Initially these landforms were thought to have resulted from the collapse of the roof of underground caverns, but this theory has been

disputed due to the absence of rock debris. It is now widely accepted that the gorges were created during **interglacials**, periods when the ice melted each summer, sending huge volumes of water cascading over the **permafrost** or permanently frozen ground, and causing rapid surface erosion.

Although in some places there is a thin cover of **glacial drift** (the boulder clay deposited over the limestone by glaciers), the soil cover is usually thin, with much of the area grass-covered moorland. This thin soil cover is easily eroded, and in places the underlying rock has been exposed as a **limestone pavement**, probably the most distinctive landscape feature of glaciokarst (Figure 1.34). Here the well developed rock joint system of Carboniferous limestone is clearly revealed as percolating rainwater enlarges the joints or **grykes**, which separate blocks of limestone called **clints**. Many limestone pavements were exposed during the last glaciation and have subsequently been eroded and dissolved by rainwater. Some pavements, e.g. Malham Cove were formed beneath boulder clay. Water, carbon dioxide and bacteria in the soil eroded the underlying limestone, which was then exposed when the ice advanced and removed the thin surface cover. Close examination of Figure 1.34 reveals the presence of small runnels cut into the surface of the clints, which channel rainwater and contribute to the chemical breakdown of the rock surface by solution. In addition to erosion by water and frost action, the pavements are also subject to biological action, because the grykes contain rock fragments and soil which provide a shaded and humid habitat for flora and insects.

Figure 1.34 Limestone pavement in the Yorkshire Dales National Park. C

Another common feature in areas of karst scenery are enclosed surface depressions known as **dolines**, or in the Yorkshire Dales, as **shake holes**. They vary considerably in size ranging from one to several hundred metres in diameter. They are formed when the thin overburden of soil is slowly flushed down through an open fissure in the underlying limestone, causing the surface to sag and subside. Chemical weathering leads to mass wasting of the sloping surfaces which deepens the hollow. Large areas of Malham Moor are pockmarked with numerous shake holes which have developed in the thin covering of boulder clay deposited by the retreating glaciers after the last ice age.

The harder and more resistant strata within the limestone may be exposed by erosion on the side of hills and appear as prominent cliff-like walls or **limestone scars** (Figure 1.33). Such features are also found underground in limestone areas. Where surface water flows across impervious cap rocks (such as shale) it disappears underground through a **swallow hole** or **sink** where it passes on to the jointed limestone. Stream sinks may be either narrow fissures or huge shafts, but this natural drainage flowing down through the rock creates vast networks of **potholes**, caves and underground passageways which are typical features of the solutional erosion of mountain limestone (Figure 1.29). Almost every hill in the Yorkshire Dales is honeycombed with cave and passageway systems, as rapid stream flow along the horizontal bedding planes and vertical joints in the limestone has eroded out lower levels. The upper

caves are often left drier and this has allowed the development of **stalactites** and **stalagmites** through slower percolation flow of lime-rich water through the rock.

The karst scenery of the Yorkshire Dales is of major importance in both a UK and a European context. The processes of weathering and erosion have produced a landscape which is unique in the UK; limestone pavements, cliffs and scars, cave systems and underground rivers. Associated with these features are distinctive grasslands and old woodlands. Over 35% of the limestone dales have been designated as **Sites of Special Scientific Interest (SSSIs)** by **English Nature**, the body which promotes nature conservation in England. As with so many other areas of high landscape value, the Yorkshire Dales face increased pressure from modern agriculture, quarrying and recreation.

Questions

1 Explain how the processes of glaciation, weathering and erosion have influenced the development of Carboniferous limestone landscapes.

2 List the main landscape features shown in Figure 1.29 and explain their formation.

Rural landscapes and settlement patterns

Changing farms

As in the chalk downlands of Sussex, farming has transformed the landscape of the Yorkshire Dales. The farmland of the Dales (Figure 1.35) is characterised by an intricate pattern of dry stone walls, several hundred isolated stone barns, and flower-rich hay meadows. Livestock farming is the mainstay of the local economy, mainly sheep and beef cattle, with some dairying in the lower dales. The traditional farmed landscape which has resulted is perhaps one of the most distinctive in Western Europe, but has been under threat for some years now due to pressures on farming practices (Figure 1.36a).

Farmers have to respond to changes in their industry, but some aspects of modern farming systems cause concern because of their possible long-term environmental and economic impacts (Figure 1.36b). More intensive production methods have resulted in a loss of traditional farming landscapes and have had a severe impact on flora and birdlife. With changing practices, features such as barns and drystone walls are no longer essential in the farming landscape. Traditionally, cattle were over-wintered in these barns, and their accumulated mulch was then spread on the fields as fertiliser during the spring. Hay cropped from the fields was stored in the barns and fed to the cattle during the following winter. The National Park Authority has introduced a scheme to provide funds for the repair of barns and walls in certain areas of the Park. The scheme is jointly funded by English Heritage and the EU. Some barns have been adapted for use by local craftsmen and others refurbished as visitor accommodation. Some land not suitable for livestock has been given over to new uses such as grouse rearing and farm woodlands. Many Dales farmers have already diversified into providing services for some of the 8.3 million visitor days spent in the Dales each year. Farmhouse accommodation, camp sites and farm shops have been developed to provide additional income, although much of this is seasonal.

The economy of rural areas, once strongly linked to agriculture, is also experiencing great change. Links between farmers and the local community have been weakened as mechanisation has reduced the need for local labour. This has reduced employment opportunities in rural areas and alternative jobs are often difficult to find.

The Yorkshire Dales National Park has the UK's highest concentration of herb-rich meadowland, which is generally regarded as being the habitat most vulnerable to agricultural change. Estimates suggest that 95% of hay meadows in England and Wales have been wiped out since the Second World War. Many of those which remain are now protected as SSSIs. In 1987 the Pennine Dales ESA (Figure 1.13) was designated, covering 16% of the National Park. Under this scheme five-year agreements are offered to farmers willing to manage their land in a traditional way. In return for a payment per hectare, participating farmers agree to maintain existing walls and barns, not to plough up grassland, and to keep stock off meadows for agreed periods. The use of fertilisers, pesticides and herbicides is also strictly controlled.

Figure 1.35   Traditional Dales farmland. Ⓒ

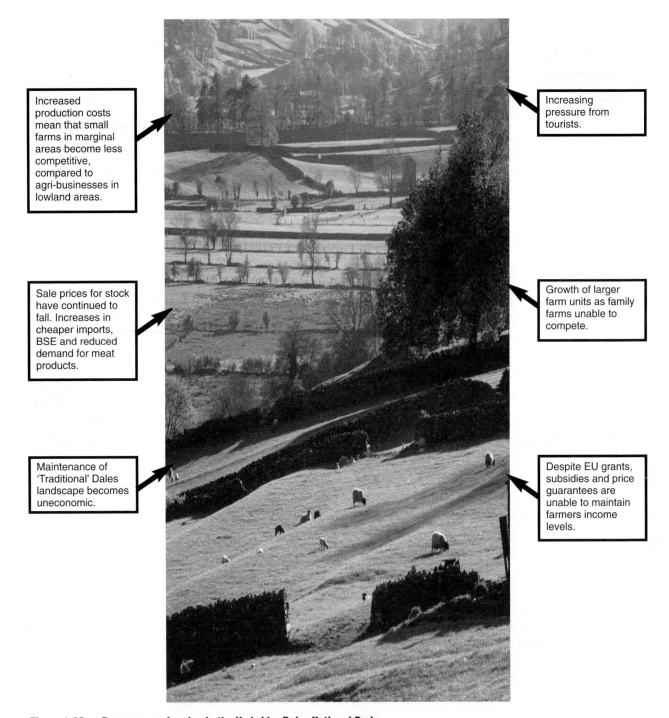

Increased production costs mean that small farms in marginal areas become less competitive, compared to agri-businesses in lowland areas.

Increasing pressure from tourists.

Sale prices for stock have continued to fall. Increases in cheaper imports, BSE and reduced demand for meat products.

Growth of larger farm units as family farms unable to compete.

Maintenance of 'Traditional' Dales landscape becomes uneconomic.

Despite EU grants, subsidies and price guarantees are unable to maintain farmers income levels.

Figure 1.36a   Pressures on farming in the Yorkshire Dales National Park.

The changing village

The basic pattern of settlement in the Yorkshire Dales was established by the Danish farmers who colonised the main valleys between the seventh and tenth centuries. By the time of the Domesday Survey of 1086 there were few nucleated villages, but Norman society encouraged village growth, and the present pattern of nucleated villages was well established by the Middle Ages.

These villages, built of local material and in a traditional local style are fundamental to the Dales landscape. Throughout the UK, villages were largely created to fulfil an agricultural function which has become increasingly less relevant. As the agricultural labour force has declined, and with it the population of rural areas, village services have followed this decline. Over the years, villages have lost their schools, post offices, village shops and cottage

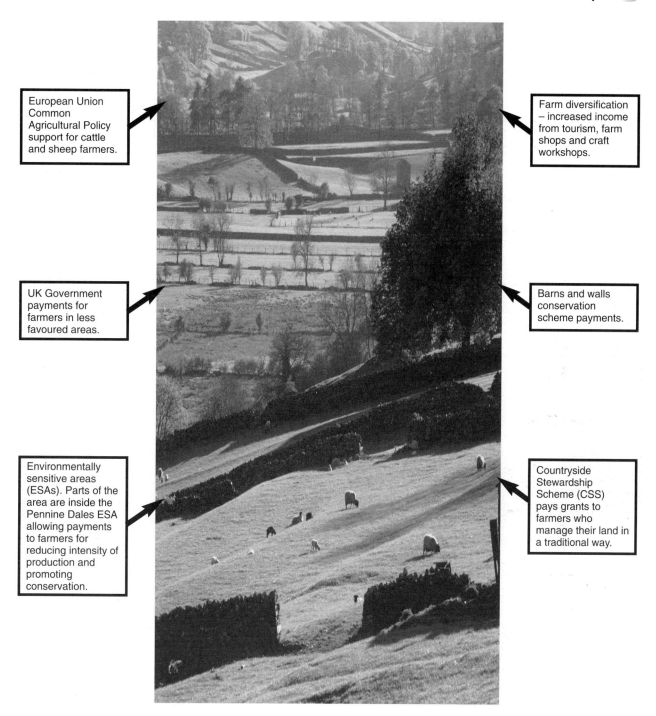

European Union Common Agricultural Policy support for cattle and sheep farmers.

Farm diversification – increased income from tourism, farm shops and craft workshops.

UK Government payments for farmers in less favoured areas.

Barns and walls conservation scheme payments.

Environmentally sensitive areas (ESAs). Parts of the area are inside the Pennine Dales ESA allowing payments to farmers for reducing intensity of production and promoting conservation.

Countryside Stewardship Scheme (CSS) pays grants to farmers who manage their land in a traditional way.

Figure 1.36b   Responses to pressures.

hospitals, leaving people to travel long distances for even the most basic services. A Survey of Rural Services was carried out in 1997 for the parliamentary Select Committee on Agriculture. This survey revealed that for many of the basic services required for the conduct of normal daily life, rural parishes were often very much worse off than even the most deprived urban areas (Table 1.1).

Despite the picture of rural deprivation revealed by Table 1.1. Figure 1.37 indicates that between 1984 and 1998 the population of rural districts in England grew appreciably faster (10.3%) than that of England as a whole (5.3%). This increase is primarily due to a net in-migration to rural areas, mainly of relatively wealthy people from towns and cities. At the same time, out-migration from rural areas has continued, mostly of the younger generation to find work and homes that they can afford.

Service	% of rural parishes without service
Permanent shop (of any kind)	42
General store	70
Post office	43
Bank or building society	91
Bank cashpoint machine	95
Petrol station	56
Newsagent or confectioner	86
School (for six year old)	50
School (for any age)	49
Public nursery	93
Private nursery	86
GP (based in the parish)	83
Dentist	91
Pharmacy	79
Village hall/community centre	28
Public house	29
Library (permanent or mobile)	12
Sports field	50
Youth club	68
Daycare group for the elderly	91
Residential care for the elderly	80
Job centre	99
Official jobs noticeboard	98
Benefits Agency office	99
Citizens Advice Bureau	94
Daily bus service	75
Community minibus/social car scheme	79
Rail service	93
Police station	92
Public telephone	9
Mains sewerage	20
Mains gas	54

SOURCE: 1997 SURVEY OF RURAL SERVICES,
SELECT COMMITTEE ON AGRICULTURE

Table 1.1   Proportion of rural parishes in England without services (1997).

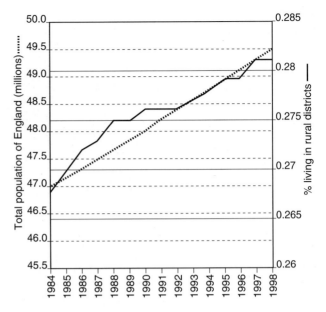

Figure 1.37   Rural population change.

One consequence is that people in rural areas are, on average, slightly older than their urban counterparts (Figure 1.38). Recent studies carried out for the Countryside Agency have identified a number of threats to community spirit and social cohesion in rural areas:

- increased mobility and rapid population change
- more people commuting instead of working in the community
- decline of local shared facilities such as shops, schools and pubs
- loss of distinctive rural culture, language and dialects
- conflicts between incomers, locals and farmers
- loss of younger people.

Rural communities within the Yorkshire Dales National Park exemplify many of these problems. Wensleydale (Figure 1.31) in the north eastern part of the Park exhibits such symptoms of rural deprivation; an aging population, lack of employment, lack of housing opportunities for local people, and declining village services. In an attempt to address these problems, the National Park Authority has established a number of initiatives. In Hawes, a small industrial site was established in association with the former Rural Development Commission (now part of the Countryside Agency) and English Estates. Seven small industrial units were built on a local authority site in the mid 1980s and by 1989, this had increased to ten units.

Although small scale, these small firms can help to offset the continuing decline of farm-related employment, allowing local people to work close to where they live. With the continuing trend towards **footloose industries** (without locational ties) and high-technology based light industry, rural areas offer many advantages to potential employers, including a stable work force and a high quality of life.

Provision of jobs and services alone will not solve the problems in rural areas. Low cost housing is also vital if local families are to be encouraged to stay in the villages. West Burton (Figure 1.39) near Aysgarth, is in an upland area and was originally built around a ford and a crossroads with the local economy based on lead mining and agriculture. In a recent survey almost half the village population of 270 were in the 'dependent' category, with 60 aged under 16 and 70 of pensionable age. The village is reasonably well provided with services. There is a post office, village shop, butcher's shop, mobile library (which calls every three weeks) and a weekly visit from a fish and chip van. The nearest dispensing doctor's surgery is in Aysgarth, 3 km away although there are regular prescription deliveries to the local post office. There is a pub, and the village primary school has 34 pupils travelling in from a wide area around West Burton. Buses run most days to both Leyburn and Hawes and there is an informal car-sharing network.

Figure 1.39   West Burton, Wensleydale.

Most of the working population commute to jobs beyond the village, although up to 20 work on family farms and about ten in local service industries. A number of jobs are tourism-related, i.e. part-time and seasonal in nature.

The National Park Authority is particularly concerned that a substantial number of the area's housing stock is not occupied full-time. Of the 90 houses in West Burton, 32 are second or holiday homes. Eleven houses are privately rented and permanently occupied, and both of the original council houses were sold to their tenants. The local community need more affordable housing to be built in the village.

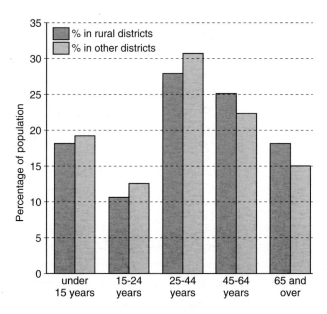

Figure 1.38   Rural population age structure.

Questions

3 Identify and explain the main differences in farming patterns in the Yorkshire Dales and the South Downs.

4 Why is 'the calendar image of a timeless, unchanging Dales village' both outdated and inaccurate?

5 Explain why second homes are often seen as a mixed blessing in rural areas.

Conflicts in the park

Quarrying in the Yorkshire Dales

The landscape character of the Yorkshire Dales is largely the result of the underlying geology (Figure 1.32). In the same way that this landscape is a valuable resource for the development of tourism, so the rocks and minerals are valuable raw materials. One of the problems for the National Park Authority is to attempt to reconcile the fact that whilst they are charged with protecting the distinctive Dales landscape, they also have to ensure that this aim is balanced with the employment opportunities which activities such as quarrying provide for local communities. Quarrying has traditionally been an important source of local employment and whilst modern large scale mineral extraction employs fewer people, it is estimated that the quarrying industry provides employment for 7% of the Dales' working population, as well as contributing almost £8 million a year to the local economy.

At present two main rocks are quarried within the National Park, **gritstone** and limestone. Gritstone is derived from Ordovician and Silurian rocks, mainly siltstone and sandstone and is used for a variety of surfacing operations, notably roads and airport runways, because of its ability to resist skidding. Carboniferous Limestone has a high content of calcium carbonate and is used for agricultural and industrial lime, flux for the steel making industry, glass making and for making toothpaste.

The Yorkshire Dales National Park has the greatest concentration of limestone pavement in the UK (Figure 1.40) with the areas around Malham and Arncliffe providing the most extensive exposures. Limestone in this area has proved a popular resource since Prehistoric times when clints were removed and used to build stone circles. During the medieval period, intensive grazing by animals caused soil erosion and exposed further areas of limestone which was burned in the many lime kilns found throughout the Dales. The ash was used to improve the poor grazing land. Later, during the enclosure periods of the eighteenth and nineteenth centuries the stone was used for walls to separate the fields. Recently, a new threat has emerged with increased demand for limestone as a result of the flue-gas desulphurisation programme of the electricity generating authorities.

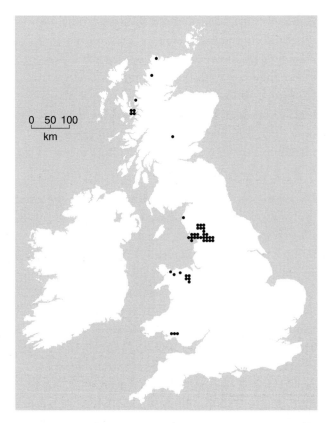

Figure 1.40   The distribution of limestone pavement in the UK.

Figure 1.41 shows the major quarries still working within the National Park area. The concentration of quarrying activity in the south-west part of the Park on the Kilmsey and Malham or Great Scar Limestone (Figure 1.32) is due to the rock's high chemical purity, with calcium carbonate content generally in excess of 95%. The limestone is several hundred metres thick, and it is relatively easy to quarry. There are fewer quarries today than in the past, but those that remain are much larger and the environmental issues associated with their operation considerable. Their size creates a visual intrusion on the landscape (Figure 1.42) because the very scale of operation makes concealment impossible. Continued activity also threatens landscape features such as cave systems and wildlife. Blasting, crushing and screening of rock, the loading and burning of lime and transport to and from the quarries create noise and dust emissions which can often be a source of annoyance to both residents and visitors. Transportation of the limestone causes particular problems with heavy traffic concentrated on certain minor roads.

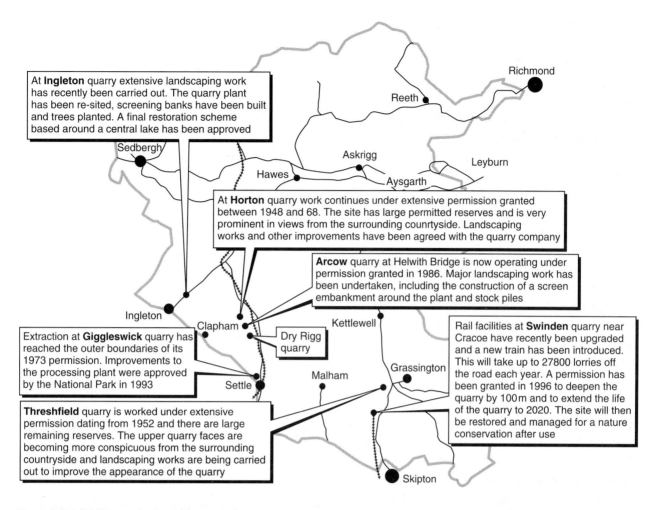

Figure 1.41 Working quarries within the Yorkshire Dales National Park.

Tourist authorities have suggested restricting such traffic but hauliers are keen to increase the maximum gross weight allowed and thereby allow more economic transport of quarry products.

Swinden quarry

Swinden Quarry (Figure 1.43) has been worked for limestone extraction and processing since 1900. In 1996 planning permission was granted to deepen the quarry by 100 m and to extend the working life of the operation up to the year 2020. Although planning permission had been refused for a previous application in 1993, the alternative proposal presented by the quarry owners Tilcon in 1996 addressed many of the concerns of both local residents and the planning authority:

'To meet the higher environmental standards desired in today's world, Tilcon is in the process of implementing environmental management systems throughout its operations with Swinden Quarry being one of the first'.

Figure 1.42 Visual intrusion of a quarry.

Figure 1.43   Swinden Quarry.

Figure 1.44   Swinden Quarry after restoration. ⓒ

Landscape

- The majority of the processing plant, which previously caused a severe visual impact has been moved to a new location 20 m below the existing quarry floor and entirely within the quarry (Figure 1.44).
- Mineral extraction will be confined within the existing quarry boundaries and the remaining landscape of Swinden Hill retained by quarrying to a depth of 100 m from beneath the remainder of the quarry floor.

Transport

- Tilcon has recently upgraded rail facilities at the quarry and a new train has been introduced. It is estimated that this will result in a 23% decrease in heavy lorry movements; 173 fewer HGV movements every day.
- Road haulage will still account for most movement of quarried material and the company insist on several environmental measures including the compulsory 'sheeting' of all vehicles. (That is, the covering of all lorries carrying quarried stone with tarpaulins, so that dust and rocks do not escape.)

Environment and community

- The extension of the existing mineral waste disposal area will be progressively landscaped to reinforce the northern slopes of the hill.
- At the end of its working life the quarry will be restored and a nature reserve created, with a lake occupying the site.
- A liaison group has been created in association with the local parish council to ensure that any concerns about the operation of the quarry can be addressed.

Recreation and leisure in the Yorkshire Dales

Almost ten million people live within 90 minutes driving time of the Yorkshire Dales National Park (Figure 1.45). It is estimated that over ten million day visitors use the Park every year, with the numbers reaching a peak during the summer months (Figure 1.46). In the southern part of the Park, day visitors are particularly important reflecting the proximity and ease of access from the conurbations of West Yorkshire, Greater Manchester and Merseyside (Figure 1.47).

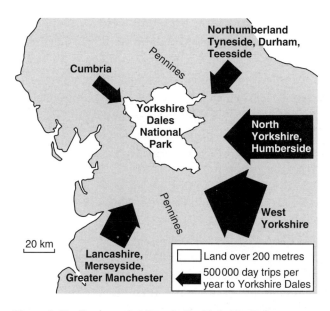

Figure 1.45   Sources of visitors to the Yorkshire Dales National Park.

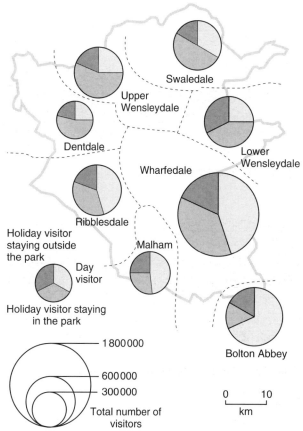

Figure 1.47   Distribution of visitors to the Yorkshire Dales National Park.

The local economy has benefited substantially as visitors create increased employment and income, especially since a wide range of activities are spread across the park (Figure 1.48). A visitor study carried out during the 1990s suggested that spending by Park visitors amounted to over £45 million annually and directly supported 1500 jobs (Table 1.2).

The increasing number of visitors is now placing a considerable strain on certain popular areas within the park and is causing environmental problems which include:

- severe footpath erosion
- disturbance to farm livestock and wildlife
- traffic congestion on certain routes at peak times
- parking problems at visitor attractions, leading to damage to village greens
- increased commercialisation of traditional villages
- pressure for new developments to provide for and to attract visitors.

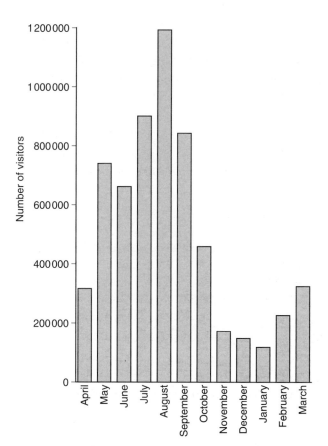

Figure 1.46   Monthly visitors to the Yorkshire Dales National Park.

Job type	Numbers of jobs
Food shops	50
Specialist shops	239
Cafes/pubs	446
Attractions	32
Garages	80
Public transport	2
Parking	37
Serviced accommodation	400
Self-catering	452
Hostels/bars	60

SOURCE: YORKSHIRE DALES VISITOR STUDY 1991

Table 1.2 Tourism generated jobs in the Dales.

Some areas within the park attract a lot of people and these 'honeypot' sites need to be sensitively managed. One such site is the Malham area (Figure 1.49). Few other places in the UK possess upland limestone scenery on such an impressive scale. The importance of the area has recently been recognised by designation by English Nature as an SSSI and the area around Malham Tarn as a **National Nature Reserve (NNR).**

This area has been popular with visitors for over 200 years, and although Malham village had a population of only 134 at the last census, the huge numbers of visitors have helped to maintain local services, including hotels, cafes, shops and pubs which might otherwise have closed. At certain times during the year the levels of traffic and car parking can substantially reduce the appeal of the village (Figure 1.50). Official car parks quickly fill up and vehicles park on road sides and in passing places on the narrow roads.

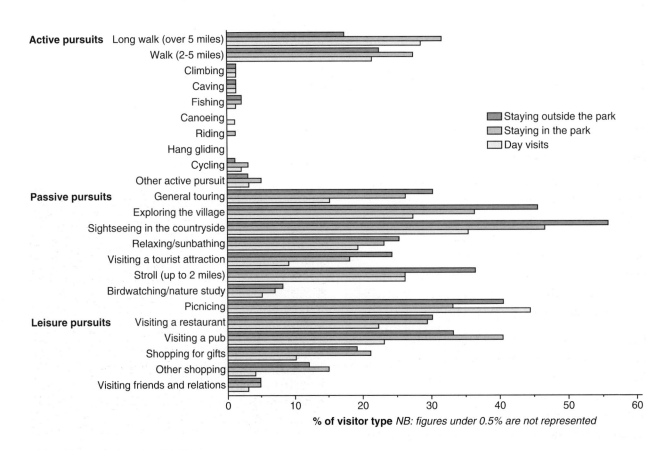

Figure 1.48   Activities by visitor type.

Malham Cove
This great limestone amphitheatre was formed when ice and water eroded the hillside back from the line of the Mid-Craven Fault. At times it must have been a spectacular waterfall, as abundant meltwater flowing down the Watlowes valley in the ice age tumbled 70 metres into the ice-filled Cove. Obviously, with no surface flow across the limestone plateau today, the Cove remains dry, except in times of severe flooding (such as 1969). The Cove face is one of the few natural nest sites for Housemartins, and is a refuge for rare plants like the Pennine Whitebeam

Malham Tarn
This attractive lake lies on a bedrock of slates in a depression scoured out by glacier ice in the ice age. The occurence of Silurian slate, surrounded on all sides by younger Carboniferous Limestone, is the result of earth movements and erosion in the geological past. The Tarn is scientifically important because springs, rising from the surrounding hills, bring in dissolved limestone, making it the highest lime-rich lake in the country. The unusual habitat of the Tarn is therefore a home for a unique collection of plants and animals

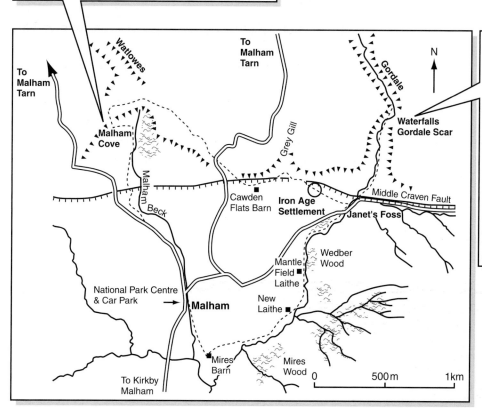

Gordale Scar
Though on the same scale as Malham Cove, Gordale Scar is dramatically different. It is a twisting gorge, eroded through the thickly bedded limestone by meltwater streams in the ice age. The modern stream is much smaller than former ones but is notable because it is rich in dissolved limestone. This precipitates throughout the length of the gorge to form large deposits of yellowish tufa. Another tufa deposit may be observed further downstream at Janet's Foss

Figure 1.49   The Malham area.

An experimental traffic management scheme during the early 1990s was abandoned due to pressure from local people and traders who reported loss of income during the period of the experiment. There are no plans to extend car parking in the village because the existing car park is often empty on weekdays in winter. During the summer peaks, a local landowner provides additional car parking space.

Many visitors use the local services but they also put pressure on landscape resources. Visitors walking to Malham Tarn and the Cove cause footpath erosion and twenty years ago it was necessary to surface the main path to the Cove to combat the excessive erosion. Since then most of the other popular paths in the area have been surfaced.

Figure 1.50   Malham Village in Peak Season.

Even the limestone pavements have been subject to degradation by removal of stone and wear from visitors' feet and are now protected by local bye-laws, and regularly monitored by wardens. Although litter is a constant problem and a risk to livestock and wildlife, there is a policy not to provide litter bins in the area and to encourage visitors to remove their own litter. The litter problem has been substantially reduced but it is still necessary to organise litter picks in the most popular visitor areas.

Questions

6 What are the main problems associated with limestone quarrying in the National Park? How have Tilcon addressed these issues at Swinden Quarry?

7 Describe and explain:
a the monthly distribution (Figure 1.46) and
b the spatial distribution (Figure 1.47) of visitors to the Yorkshire Dales.

8 a What particular problems does the management of 'honeypot sites' such as Malham pose for planners?
b What efforts have been made to overcome these problems?

Water supply: Grimwith Reservoir

Grimwith Reservoir is located 7 km east of Grassington in the south of the Yorkshire Dales National Park (Figure 1.31). The original reservoir was not used to supply drinking water; its purpose was to compensate the River Wharfe for waters removed by dams further up its course. The new reservoir, 150 hectares in area, is over three times as large as the old one. It is used to supply the urban centres of West Yorkshire with drinking water by using the Wharfe as a 'pipeline' for part of the route. In view of the limited existing opportunities for water-based recreation in the Yorkshire Dales National Park, the new reservoir offers considerable potential for recreation development as a water sports centre. Significantly, Harrogate, Leeds and Bradford are all within 25–40 km and the site is easily accessible to this large population. The reservoir also lies within a popular recreational area, being close to Bolton Abbey and Ilkley Moor.

Figure 1.51   Grimwith Reservoir.

Apart from its relative accessibility, as Figure 1.51 reveals, the other major asset of Grimwith is its location within a high quality landscape, and the appeal of an expanse of water set within a wild moorland area.

The first signs of increased visitor use came almost as soon as the reservoir was increased in size. Yorkshire Water, the owners of the site, improved road access to Grimwith and provided a small car park, toilets, and waymarked footpaths around the reservoir. An old quarry site to the south of the reservoir was utilised for building material and infill for the dam. Since completion of the new reservoir, pressures have increased from sailing groups, walkers, fishermen and conservationist groups whose interests are not always compatible. In deciding on the type and extent of recreational development at Grimwith, the National Park Authority have to ensure that the character and appearance of the area are not damaged by inappropriate development which would have an adverse impact on the environment. Several options are available (Figure 1.52):

Option A Designation as a National Nature Reserve
The Grimwith site is of regional importance for breeding waders and wildfowl, in particular Widgeon and Ringed Plover. During the 1990s, research by English Nature indicated that the breeding success of these birds was being severely reduced by disturbance from reservoir users. The main breeding area is shown on Figure 1.52 to the west of line A-B. The other main nesting area used by these birds is within the quarry which is also used as a car park and launching site for boats and sailboards.

It is proposed that:

- the reservoir west of line C–D will not be used for any water sport activity
- the area 200 m from the shore in this part of the lake will be designated as a nature reserve
- use of the quarry area for sporting activities on the lake will be severely restricted, with no water-borne activity during March–April, and no powered craft allowed on the reserve except for safety or maintenance
- fishing will be restricted to the areas shown on the map
- the quarry area will be landscaped and its habitat capability increased for the local bird population
- entry to the reserve will be at point C, and use throughout the year will be monitored by a counter.

Option B Integrated reservoir use

The reservoir is used for water recreation by a number of local groups, most significantly by Grimwith Sailing Centre. The proposal recognises the conservation needs outlined in Option A, but suggests the development of the Grimwith site as a multi-purpose reservoir by proposing a plan which aims to meet the demands of the differing interest groups.

It is proposed that:

- a further changing block will be built near to the existing toilets, this will include a first aid post and be funded by watersport users
- the lake shore near the quarry area will be reinforced with concrete to make transport to the area easier
- all sailing craft will be brought to, and taken away from, the area after each visit, and a log kept of all water users
- windsurfing, sailing and canoeing will be allowed throughout the year; use of power boats will be restricted to safety purposes
- the area to be used for water sports will be to the east of line A–B only
- monitoring of the breeding bird population will be undertaken
- a committee of representatives from interested groups (Yorkshire Water, Pately Bridge Outdoor Centre, Grimwith Sailing Club, The Ramblers Association, English Nature, Yorkshire Dales Society, Yorkshire Dales Upland Bird Study Group, Yorkshire Dales National Park Authority, and an angling representative) will meet twice annually to discuss any problems arising from the use of the reservoir and agree action.

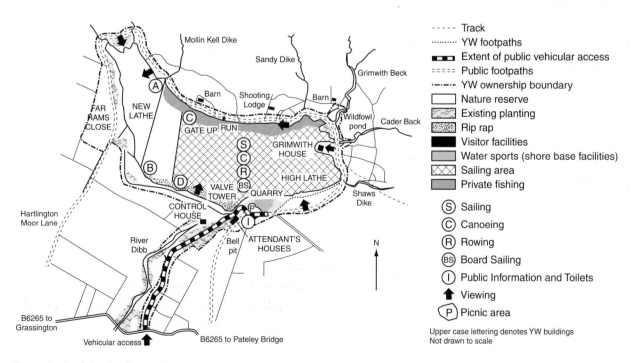

Figure 1.52 Grimwith Reservoir.

Option C The Grimwith Recreation centre

If the use of Grimwith Reservoir for watersports is to continue and be commercially viable, it is suggested that a medium scale outdoor pursuits centre is needed on the site.

It is proposed that:

- a water sports and outdoor pursuits centre be built, comprising: a clubhouse with changing facilities, showers and a toilet block; an accommodation block for 30 people; a warden's house.

All buildings will be constructed from local stone or similar materials with extensive landscaping, particularly tree planting. Trees will also be planted at strategic locations around the reservoir to screen areas used by the various bird species in order to minimise disturbance from walkers using the footpaths

- provision will be made to use the centre for environmental education as well as outdoor pursuits
- a slipway and small jetty will be built into the lake from the existing car park
- a new car park will be made inside the quarry area and provision made for a camping area
- an alternative breeding site to replace the quarry will be made on the promontory adjacent to Grimwith House with specifications agreed with RSPB and English Nature.

Decision making

Examine the three options for Grimwith Reservoir carefully. In making a planning decision in favour of, or against, any of the options, the Yorkshire Dales National Park Authority is guided by policies outlined in its local plan, adopted in 1966. These include:

(VF1) Proposals for visitor developments will be considered acceptable if they:

(i) provide facilities or services that support informal recreation; or

(ii) bring a material conservation gain; or

(iii) assist in visitor management

In addition, proposals should:

(a) have no adverse impact on the landscape, built environment, wildlife or archaeology of the National Park;

(b) be in scale and in keeping with the character of their surroundings;

(c) not result in noisy, intensive or intrusive activities;

(d) not be detrimental, cumulatively, to the interests specified in (a), or to the local community, by reason of adding to existing visitor facilities and attractions;

(e) have satisfactory road access;

(d) display a high standard of design and respect for the building traditions of the locality.

(LC12) Development proposals that are likely to adversely affect rare, endangered or other species of acknowledged conservation importance will not be permitted.

These are underlined by the twin aims of National Parks as outlined in the 1995 Environment Act (p21), and if there is a conflict between these two purposes, greater weight should be given to conservation issues.

For each of the three options, A-C, draw up a list of arguments for and against the proposal using the planning policies listed above. On the basis of this analysis, make a reasoned recommendation as to which option should be approved.

1.4 Glaciated landscapes

A great deal of publicity has been given in recent years to the concept of **global warming**. It is widely thought that the world is currently entering a warm period, with average temperatures increasing and sea levels rising due to the melting of ice caps. Such a change is part of a cyclical process. Research has revealed that if solar radiation is reduced over a long period of time (causing lower average temperatures) a glaciation will take place.

During the past 2.5 million years there have been periods of intense cold when snow accumulated in the mountains of Scotland, northern England, Wales and Ireland and formed **glaciers**. During extended periods of cold climate, these glaciers joined up to form vast ice sheets. In between these cold periods, during **interglacials**, the glaciers began to disperse from these centres of ice accumulation (Figure 1.53). As they moved outwards their effect on the landscape below the ice was dramatic and the Scottish Highlands, the Lake District and Snowdonia were all etched with the characteristic features of glaciated upland landscapes (Figure 1.54).

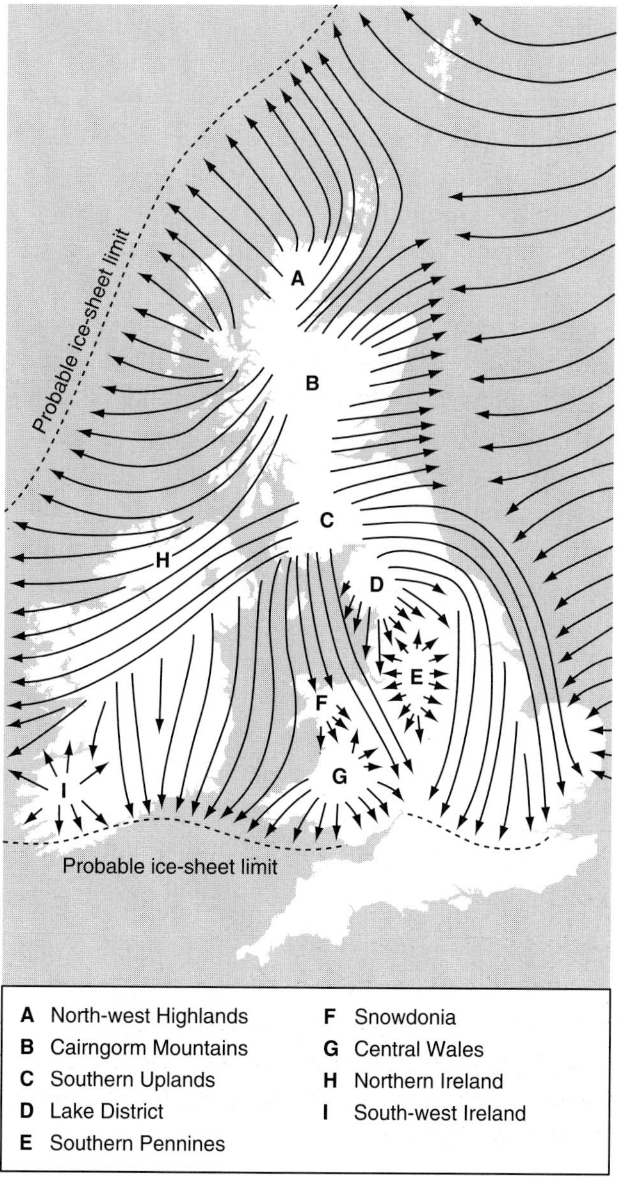

Glaciation in the UK

During the ice ages, the extreme southern part of the UK, south of the River Thames and the Bristol Channel, was not glaciated (Figure 1.53). This region experienced a tundra–like climate, similar to areas such as northern Canada and Alaska today. For most of the year the ground was permanently frozen **permafrost**. These areas were subject to seasonal surface melting of the frozen ground, and this water was swollen by the meltwaters draining from the ice sheets further north. In chalk areas such as the North and South Downs, this **fluvioglacial erosion** by the meltwater is thought to be responsible for the formation of the characteristic dry chalk valleys.

A	North-west Highlands	F	Snowdonia
B	Cairngorm Mountains	G	Central Wales
C	Southern Uplands	H	Northern Ireland
D	Lake District	I	South-west Ireland
E	Southern Pennines		

Figure 1.53 Ice movement and centres of ice dispersal.

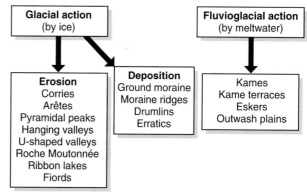

Figure 1.54 Glacial landscape features.

39

In contrast, most of the eroded sands and gravels from the meltwater were washed into the valleys of the main rivers such as the Severn and the Thames, where they were later carved into **river terraces** above their flood plains by the action of the rivers.

Glacial landscapes and processes in Scotland

Of all geomorphological systems, glaciation has had the greatest impact on Scottish scenery. Figure 1.55 shows that glaciation has been very selective in its effects, with a marked contrast evident between the eastern and western Highlands. The west consists of a rough ice-scoured landscape where the mountains have been heavily dissected by glacial troughs and valleys. The east shows much less landscape modification with evidence of erosion often confined to the valleys.

To some extent, these variations can be explained by the different glacial histories of the east and west of the country. The west had a maritime climate and glaciers built up first in these environments and remained there for longer. In addition, there were periods such as the **Loch Lomond Re-advance**, 11 000 years ago when glaciers only existed in the west. It has been suggested that the greater evidence of glaciation in the west is therefore mainly due to longer exposure to glacial activity.

Glaciers may be either frozen to the underlying rocks or be able to slide over them. Erosion takes place when glaciers slide, but this will only happen if the temperature at the base of the glacier is close to melting point. When the ice base temperature is well below 0°C, the lowest ice layer is frozen on to the underlying surface and therefore much less erosion takes place. During the ice ages, the western part of Scotland was warmer and wetter, and the underlying slopes were steeper than in the east of the country. Together, these factors favoured fast-moving ice and conditions ideal for erosion. The most intense erosion took place in the west where the lochs and valleys are deepest, sometimes forming **fiords** or heavily glaciated coastal valleys. In the east, conditions were drier and colder and the slopes more gentle, resulting in much less ice movement and therefore less erosion. Due to the varied preglacial landscape, particularly the bedrock and channel system, the local effects of glaciation varied considerably, and different parts of Scotland now show different types of landforms and different intensities of erosion. These factors affecting the rate of glacial erosion are summarised in Figure 1.56. Contrasting types of glacially eroded scenery can be identified in the Cairngorm mountains and the Flow Country of Caithness and Sutherland.

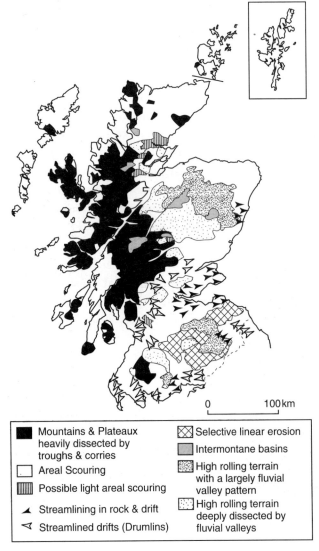

■ Mountains & Plateaux heavily dissected by troughs & corries	⊠ Selective linear erosion
□ Areal Scouring	▦ Intermontane basins
▥ Possible light areal scouring	▨ High rolling terrain with a largely fluvial valley pattern
▲ Streamlining in rock & drift	⊡ High rolling terrain deeply dissected by fluvial valleys
◁ Streamlined drifts (Drumlins)	

Figure 1.55 Glacial landscapes in Scotland.

Questions

1 Why were the upland areas of the UK the main centres for ice dispersal after the ice age?

2 Using Figures 1.55 and 1.56 describe and explain the differences in glaciated landscapes in the east and the west of the Scottish Highlands.

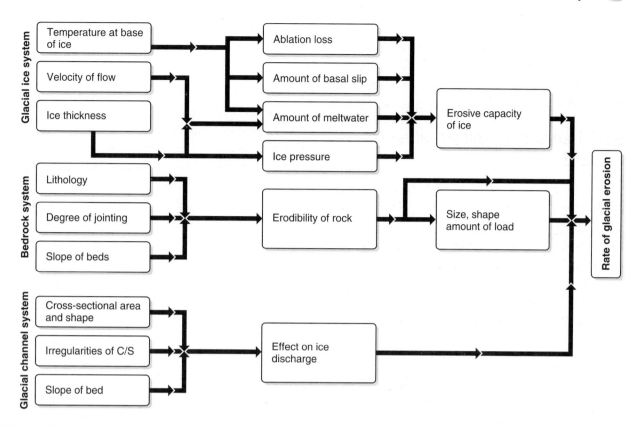

Figure 1.56 Factors affecting rates of glacial erosion.

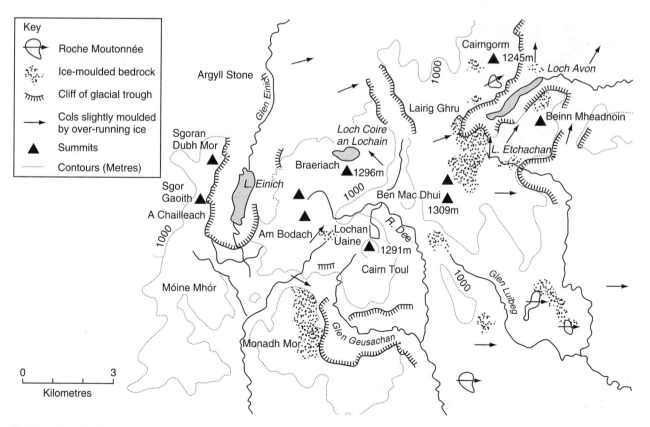

Figure 1.57 Glacial erosion in the Cairngorms.

Figure 1.58   The Cairngorm Plateau.

The Cairngorm mountains

At a local scale, the influence of geology and preexisting landforms on ice-flow during glaciation was responsible for the patterns of glacial erosion. The plateau of the Cairngorm mountains, situated between the valleys of the rivers Dee and Spey in northern Scotland, forms the largest area of land over 1000 m above sea level anywhere in the UK. Within the Cairngorms (Figure 1.57) there are four summits over 1200 m. Ben Mac Dhui (1309 m), Braeriach (1296 m), Cairn Toul (1291 m) and Cairngorm (1245 m). Between these peaks stretches an undulating granite plateau (Figure 1.58) containing classic glacial landforms such as deep troughs and massive **corries**

gouged out by glaciers during the last ice age (Figure 1.57). The ice direction, moving from west to east can be identified from the orientation of **roches moutonnées**, streamlined glacial features aligned in the direction of the former ice flow (Figure 1.59).

The plateau surface

The surface of the Cairngorm plateau is not completely flat but comprises smooth rolling slopes and shallow valleys, contrasting sharply with the deep glacial troughs cut by glaciers. The Cairngorms owe their rounded appearance to their granite composition which has weathered down to **scree** on the plateau tops by the enlargement of joints in the rock by freeze-thaw action.

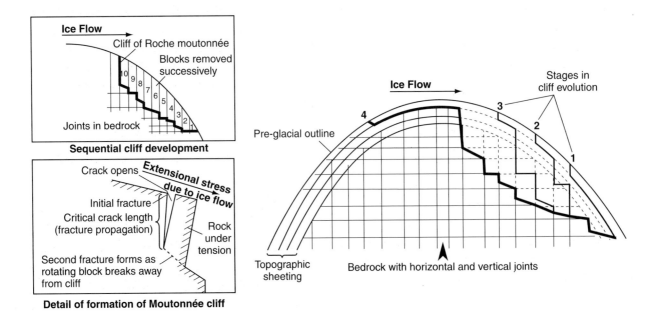

Figure 1.59   Formation of a roche moutonnée.

Figure 1.60   The Cairngorm Tors.

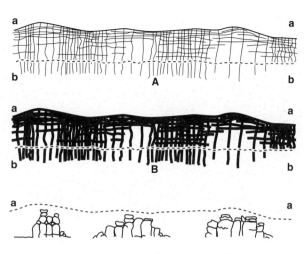

Diagram to illustrate the formation of tors. An original rock surface with variations in joint frequency (A) is attacked by chemical weathering to form areas of deeply rotted rock (black) and irregular remnants of fresh rock (B). In the bottom drawing the deeply rotted rock has been stripped by erosive processes to leave the fresh rock tars. (After Linton 1955)

Figure 1.61   The formation of tors.

Cairngorm granite has a high feldspar content and is pink in colour. It was formed from molten lavas pouring out of the Earth's crust 400 million years ago. In addition to glaciated landforms, the Cairngorms also exhibit features developed by the erosion of granite.

The Cairngorm **tors** are the finest in Scotland and consist of gigantic blocks of rocks (Figure 1.60) rising up to 25 m above the smooth hill slopes. Tors were formed before the ice ages when the climate was hot and wet. The granite was deeply weathered to an uneven depth, leaving areas of harder bedrock between areas of more heavily weathered and crumbling granite. During the ice ages the rotted granite was eroded, leaving behind some of the larger blocks (Figure 1.61). Many have since weathered into fantastic shapes influenced by rock jointing. Tors are particularly common on the eastern Cairngorm summits such as Ben Avon and the Barns of Bynack.

The plateau edges and the valley

On the edges of the Cairngorms, small mountain glaciers have cut other landforms. During the ice ages snow accumulated in armchair-shaped corries consisting of rock basins overlooked by rugged cliffs (Figure 1.62). Today these basins are often deep enough to hold small **lochans** such as Coire an Lochain to the north of Braeriach and Lochan Uaine north of Cairn Toul. An analysis of the distribution of the Cairngorm corries reveals that most face north and east, because the glaciers built up more quickly on these shady slopes where the accumulating snow was more protected from the sun's rays. The accumulated snow hardened to form ice.

Subsequent erosion through **nivation** (snowpatch erosion), freeze-thaw action, **solifluction** (soil flow), meltwater erosion and chemical processes, deepened the corries. Eventually the accumulated ice began to flow away from the steep back wall of the corrie by **rotational sliding**, further deepening the corrie as it did so (Figure 1.63). Where several corries occurred close together, they were the source for the larger valley glaciers (Figure 1.64) which excavated deep **U-shaped glacial troughs** such as the great cliffed valleys now occupied by Lochs Avon and Einich (Figure 1.65). Some valleys, such as the Lairig Ghru cut through major watersheds and became important routeways through the mountains (Figure 1.66).

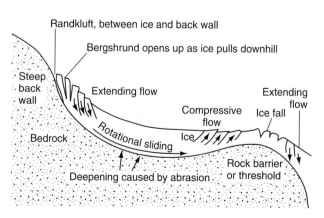

Figure 1.63   The development of a corrie.

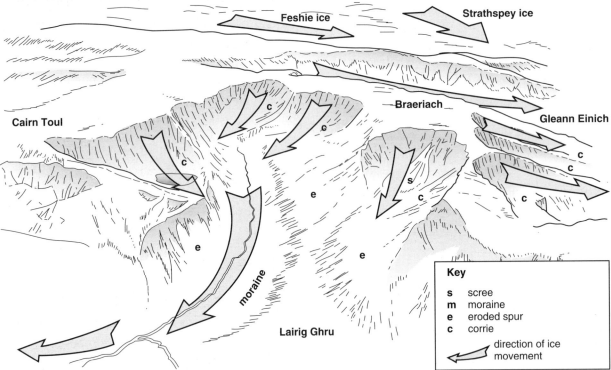

Figure 1.62   Corries of Braeriach and Cairn Toul.

When the ice began to melt, erosion became intense and created a number of fluvioglacial landforms resulting from meltwater and massive deposition of glacial debris. Much of the meltwater flowed through Strathspey where the river volume increased, particularly during the summer melt periods. The swollen rivers Feshie and Spey deposited large quantities of boulders, sand and gravel on the valley

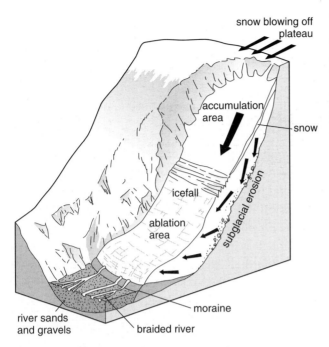

Figure 1.64   The formation of a valley glacier.

floor which have subsequently been cut into by the rivers, creating terracing and **braided river channels** where the water flows through several channels (Figure 1.67).

The Cairngorms exhibit a landscape typical of selective **linear erosion**. There was minimal erosion at the base of the ice sheet which was frozen to the bedrock for much of the time, making it relatively protective over the high ground of the Cairngorm plateau. In contrast, where the ice sheet was channelled into valleys, it was thicker, warm-based and therefore erosive. This helps to explain the differing glacial landscapes within the Cairngorms where deep glacial troughs such as that occupied by Loch Avon contrast with the plateau surface, dotted with tors and weathered rocks – features which would not have survived beneath an active ice sheet.

Q uestions

3   Using diagrams, explain the formation of both roches moutonnées and corries. For each feature, explain the significance of their spatial pattern in the Cairngorms.

4   Describe the processes involved in the formation of tors and explain their occurence on the Cairngorm Plateau.

5   Make a fully annotated copy of Figure 1.65, identifying and explaining the main features of the glaciated landscape shown on the sketch.

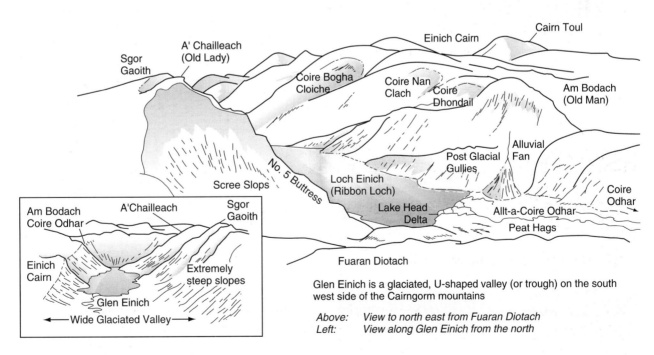

Glen Einich is a glaciated, U-shaped valley (or trough) on the south west side of the Cairngorm mountains

Above: *View to north east from Fuaran Diotach*
Left: *View along Glen Einich from the north*

Figure 1.65   Glen Einich – a glaciated valley.

Figure 1.66   The Lairig Ghru.

Conservation in the mountain areas of Scotland

A study of the map of National Parks in England and Wales (Figure 1.30) reveals that many have been designated in areas of upland mountain scenery. If the Cairngorms were in England they would also be given the protection afforded by national park status, in the same way as the English Lake District and Snowdonia in Wales.

During the 1990s it became increasingly clear that certain parts of Scotland were coming under increased visitor pressure largely due to three main factors:

- increased affluence of the population
- greater leisure time
- improvements in personal mobility.

Figure 1.68 indicates just some of the problems and pressures which tend to be concentrated at certain popular scenic locations.

Although many areas in Scotland have been designated **National Scenic Areas** (Figure 1.69), Scotland has no national parks. In 2000, primary legislation was approved by the Scottish Parliament for the establishment of the first two national parks in Scotland. It is expected that these will be established by 2002 in the Cairngorms and Loch Lomond and the Trossachs. Other likely areas for future designation include Wester Ross and Glencoe (Figure 1.70).

Designation as national parks will hopefully allow integrated and effective planning and management of these areas. There seems little doubt that the creation of national parks will be a double-edged sword; they will provide greater environmental protection but at the same time, heighten the profile of these areas which will encourage yet more visitors, thereby intensifying the existing pressures.

Figure 1.67   River Feshie.

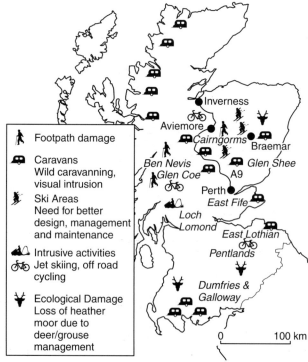

Figure 1.68   Tourism and the Scottish environment.

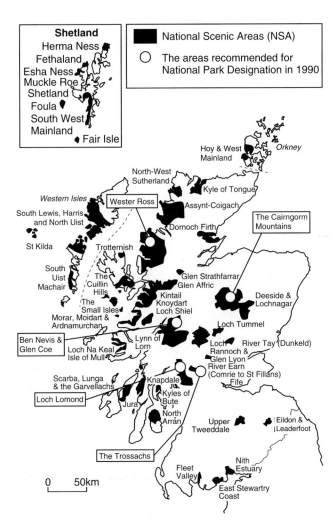

Shetland
Herma Ness
Fethaland
Esha Ness
Muckle Roe
Shetland
Foula
South West
Mainland
• Fair Isle

■ National Scenic Areas (NSA)

○ The areas recommended for
National Park Designation in 1990

Hoy & West Mainland   *Orkney*

North-West Sutherland

Western Isles   Kyle of Tongue

Wester Ross   Assynt-Coigach

South Lewis, Harris and North Uist   Dornoch Firth

St Kilda   The Cairngorm Mountains

Trotternish

South Uist Machair   The Cuillin Hills

Glen Strathfarrar Glen Affric

The Small Isles   Kintail Knoydart Loch Shiel   Deeside & Lochnagar

Morar, Moidart & Ardnamurchan   Loch Tummel

Ben Nevis & Glen Coe   Lynn of Lorn   Loch Rannoch & Glen Lyon   River Tay (Dunkeld)

Loch Na Keal Isle of Mull   River Earn (Comrie to St Fillans)

Scarba, Lunga & the Garvellachs   Knapdale   Fife

Loch Lomond   Kyles of Bute

Jura

North Arran   Upper Tweeddale   Eildon & Leaderfoot

The Trossachs

Fleet Valley   Nith Estuary

0   50km   East Stewartry Coast

Figure 1.69   Scotland: National Scenic Areas.

Questions

6 What evidence is there to suggest that certain areas of Scotland are in urgent need of stricter visitor management policies?

7 One of the landowners in Wester Ross who opposes the creation of a national park in that area has argued that it would 'create a honeypot effect'. Explain what you think he meant by this, and whether you agree.

Conservation issues in the Cairngorms

One of the most convincing arguments for the creation of a Cairngorm National Park is that despite the fact that within the area there are already 52 SSSIs and one National Nature Reserve (NNR), many conservationists argue that the protection given by these has not been enough to protect the area from what they regard as unsuitable and damaging developments.

The importance of the Cairngorms can be gauged from this assessment of their nature conservation value by Scottish Natural Heritage:

[The Cairngorms are] . . . 'more than anywhere in Britain, an authentic example of the wilderness of the northern regions of the world. This is our outpost of the boreal forests and taiga, and the tundras and fell-fields of the Arctic. It is this setting which adds so much to people's enjoyment of the animals and plants . . . the Cairngorms in their entirety are of exceptional value to nature conservation from all three of the main perspectives from which that value is assessed. First, owing to their massive extent and large altitudinal range, the Cairngorms are exceptionally rich in plant and animal species and communities, and ecological relationships. They constitute the most varied single wildlife area in Britain. Secondly, the Cairngorms show an unusual combination of oceanic and continental features, an extreme development of montane features reflecting high latitude and altitude, and a great variety of soil types in relation to parent rock type and topography. Thirdly, the Cairngorms are unique in Britain for the extent and naturalness of their arctic-alpine and boreal environments'.

The northern latitude and high altitude of the Cairngorms has inevitably had a marked influence on the local climate, which is the most continental in the UK. Braemar, on the eastern slopes is regularly the coldest place in the country during winter. There is a weather station on the summit of Cairngorm which, on average, records 196 days each year with air frost, and 83 days with temperatures below zero. Due to these severe climatic conditions, snow lies longer on these mountains than in any other part of the UK, usually for about 90 days at the summit of the chair lift at Coire Cas. This severe climate and the thin, infertile soils which have been weathered from the granite have combined to prevent even extensive farming outside the lower and more sheltered glens. As a result the Cairngorms are one of Europe's last remaining **wilderness** areas.

Flora and fauna are particularly rich in the Cairngorms, and for this reason it became the UK's first NNR in 1954. This was an attempt to preserve the unique arctic and alpine plant associations such as

a.) Wester Ross

b.) The Cairngorms

c.) Loch Lomond

d.) Glencoe

Figure 1.70   Future National Parks in Scotland.

moss campions and lichens, and the varied wildlife which includes birds such as the osprey, golden eagle, ptarmigan and snow bunting. Although it cannot be regarded as a wilderness area by international comparison with, for example the Canadian Arctic, it is nevertheless remote by most European standards, and is the largest tract of land distant from public roads in the UK. This remoteness together with the superb opportunities for winter recreation (due to the best snow-holding conditions in the UK), and summer rock climbing, hillwalking and birdwatching have combined to result in a large increase in the number of people visiting the Cairngorms. Such visitors come all year round and from an increasingly urban population. With increased accessibility via the improved A9 road, pressures on the area have increased rapidly and inevitably conflicts have developed between those keen to expand facilities for tourism, and conservation interests.

The expansion of skiing in the Cairngorms

One of the fundamental contradictions of modern countryside recreation is that as more people are encouraged to seek out remote areas, their increased numbers make a greater impact on these regions, thereby making them less attractive as wilderness areas. In order to solve this problem the concept of **sustainability** (meeting today's needs without reducing the ability to meet tomorrow's needs) has become an important consideration in many development projects.

Since the 1960s walkers, climbers and wildlife enthusiasts in the Cairngorms were joined by increasing numbers of skiers. Whereas the former groups make minimal demands on the landscape, the popularity of the Cairngorms for skiing has been accompanied by growing demands for better facilities such as new access roads, more ski runs and tours, restaurants and accommodation.

The expansion of skiing in the Cairngorms has played a vital role in the post-war local economy of the Spey Valley. Throughout the Scottish Highlands, lack of employment opportunities has frequently led to rural depopulation, particularly amongst the younger age groups. Yet in the Spey Valley,

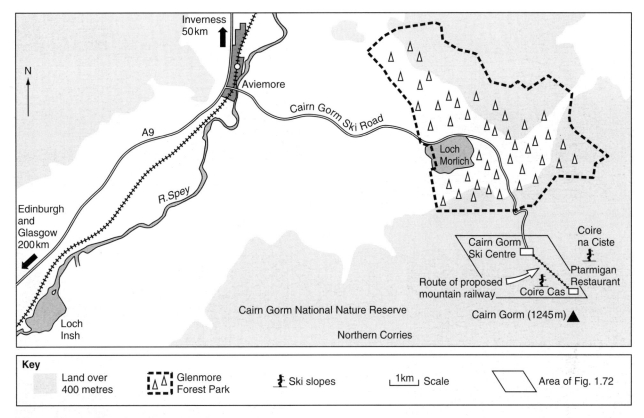

Figure 1.71 Aviemore and the Cairngorms.

tourism-related employment (40% of employees work in businesses directly related to tourism) has meant that jobs are available locally. A recent estimate suggested that skiers bring in about £12 million annually to the local economy, generating the equivalent of 350 full-time jobs.

Mechanised downhill skiing in the Cairngorms began in 1961 with the construction of the White Lady Chairlift and the extension of an access road through Glen More from Aviemore (Figure 1.71). Downhill skiing was concentrated at Coire Cas and Coire na Ciste. By 1980 these areas had been developed to their capacity, and by the mid-1980s the Cairngorm Chairlift Company produced plans for further expansion of skiing into an area known as The Northern Corries, which had previously been used principally for climbing, walking and cross-country skiing. The planning application was refused in 1990, largely after **Scottish Natural Heritage (SNH)**, the countryside advisory agency for Scotland, advised the government against the development on the grounds of visual impact, the effect on other forms of recreation and the intrusion of commercial development into an area of high and wild land.

The existing 1960s ski facilities cannot meet the increasing demands each winter. The opening of new modern facilities elsewhere in Scotland, notably at Aonach Mor in the Nevis Range near Fort William, and also the attractions of European ski resorts have had a serious economic impact on skiing in the Cairngorms. Facilities in the Cairngorms are older, unreliable and the ski lifts have to be stopped whenever the winds exceed 30 mph (and this in one of the UK's windiest areas). In 1997 the government gave its backing to a new scheme to replace the chairlifts with a funicular railway (Figure 1.72) which should create up to 150 new jobs in the area. Such a scheme will be weather-proof and capable of uplifting twice as many skiers per hour as the chairlift. The developers argue that the railway will provide a year-round tourist attraction and replace much of the existing piecemeal development with a single scheme.

Construction work on the project began in August 1999 but it has been beset with opposition arising primarily from environmentalists. In a letter to *The Times* mountaineer Chris Bonington argued that the development 'would do irreversible damage to

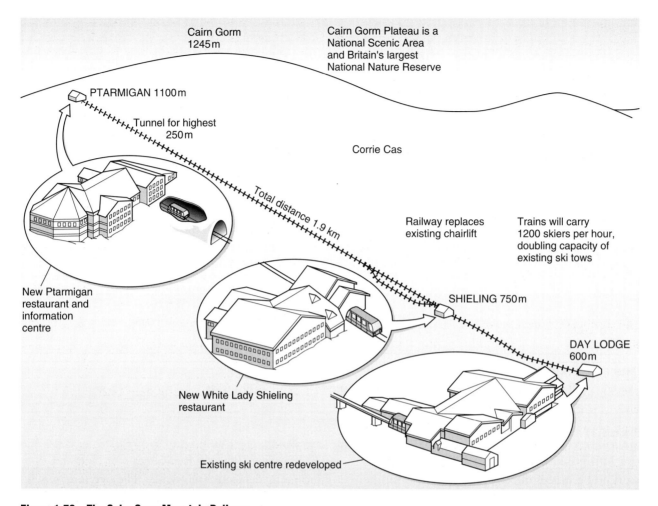

Cairn Gorm
1245m

Cairn Gorm Plateau is a
National Scenic Area
and Britain's largest
National Nature Reserve

PTARMIGAN 1100m

Tunnel for highest
250m

Corrie Cas

Total distance 1.9 km

Railway replaces
existing chairlift

Trains will carry
1200 skiers per hour,
doubling capacity of
existing ski tows

New Ptarmigan
restaurant and
information
centre

SHIELING 750m

DAY LODGE
600m

New White Lady Shieling
restaurant

Existing ski centre redeveloped

Figure 1.72   The Cairn Gorm Mountain Railway.

one of Europe's most important wild areas', and an article in *The Scotsman* went further and commented 'The Cairn Gorm funicular railway is a nonsense, a foolish scheme to rescue the ski resort from the threat of snowless winters by allowing summer visitors to damage the resource that brings people in the first place. And what a place it is, a world of sky and space, of rock and water, forest and loch. A place we should be proud of, a place we should defend and protect rather than fill with tons of concrete and install a toy train.' Developers argue that such schemes have been in operation in the Alps for many years, operating without the kind of environmental armageddon predicted by the Cairn Gorm conservationists.

Throughout the 1980s and 90s, SNH were the principal objectors to any development plan for expansion of skiing in the Cairngorms. A precondition for withdrawing their objections to the funicular plan was acceptance of a strict 'Visitor Management Plan' which would prevent casual funicular visitors from leaving the top station. They will be able to visit a cafe, visitor interpretation centre and shop, but they will not be allowed to climb to the summit of Cairn Gorm (as chairlift clients are currently able to do). Those who wish to visit the wild areas on the summits will only be able to do via the long walk in from Glenmore, i.e. by climbing the mountain.

The debate over the railway is a classic example of how polarised views between the developers and the conservationists have greatly hampered conflict resolution. As one observer commented, 'A fence was built, and everyone was forced to choose on which side they wanted to stand.' It may be that in future, dialogue and negotiation should be the basis of planning disputes such as this, the outcome of which have such profound socioeconomic and environmental consequences.

Questions

8 What physical factors are responsible for the Cairngorms being one of Europe's last wilderness areas?

9 Why did the Cairngorms become a National Nature Reserve in 1954 and what protection did this status provide?

10 Why are people in the UK increasingly attracted to remote areas for recreation?

11 Outline the arguments **for** and **against** the building of the Cairn Gorm Mountain Railway.

The changing village: Laggan and the conservation of people

The village of Laggan is situated in the upper part of the Spey Valley (Figure 1.73) at the point where the river is bridged by the A86 road which links the area with the west coast of Scotland. Two hundred years ago records indicate that the population of the parish was over 2000 people, the 1991 census put the population at 189.

The depopulation of Laggan began during the nineteenth century with the '**clearances**' as landowners evicted crofters and small tenant farmers and used the land to increase their own sheep flocks. Thousands of Highland Scots were forced to leave the area; some migrated to take jobs in the expanding industries of the cities of central Scotland while others emigrated to Canada and the USA.

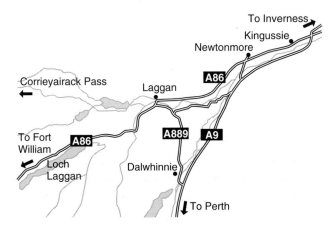

Figure 1.73 The Location of Laggan.

Laggan employment	1970	1990
Farmers	6	6
Farm employees	19	11
Estate workers	13	7
Crofters	5	4
Forestry workers	14	3
Conservationists	0	2
Arts & Crafts	0	4
Retired locals	25	11
Retired incomers	0	11
Self-employed foresters & fencers	0	6
Others living in Laggan, employed outwith Laggan	0	4

Table 1.3 Employment in Laggan, 1970–1990

Since then depopulation has continued; with increased mechanisation there are fewer jobs available on the remaining farms and estates (Table 1.3). Apart from the lack of employment, another problem for local people is the lack of affordable housing. The growth of the holiday homes market, coupled with the arrival of 'new settlers' in the form of conservationists, artists, craftsmen and retired people from the south, have all served to increase house prices and limit the amount of property available to people living and working in the area.

In Laggan the local community raised enough funds through government, European Union and development agency grants to purchase the village post office and general store (Figure 1.74) which was turned into a community cooperative venture. In addition a community hall has been built and television reception greatly improved by the locals erecting their own mast.

Changes in the village in the late twentieth century are summarised in Table 1.4. The decline in pupil numbers at the village school was a matter for concern in the 1990s, but the situation has been reversed with over half of the pupils from English

families who have moved into the area. Tourism is clearly the growth area helped by a recent popular BBC television series 'Monarch of the Glen' set in the area (Figure 1.75). This show has been sold to 12 countries and an estimated 50 million people worldwide are expected to see the programme. This has had the effect of attracting more tourists and visitors to Laggan, as well as bringing income directly to the village during the filming of the programmes. One recent survey found that as many as 9% of tourists choose to visit the Highlands because they have seen the scenery in films and television programmes. In June 2001 the local tourist board decided to capitalise on the success of the programme by renaming the area "Monarch Country", a title which will be used on brochures and road signs. It is estimated that the area could benefit by £100 million in increased tourism revenue by 2010. The dependence on tourism is therefore increasing, while traditional occupations continue to decline.

Figure 1.74   Laggan post office and general store.

1970	1990	2000 changes
School 35 pupils, 2 teachers	**School** 16 pupils, 1 teacher, 1 part-time teacher	**School** 29 pupils (5 in Nursery), 2 teachers, 1 assistant, 1 auxiliary, peripatetics
Church	Church	
Shop & petrol pumps (privately owned)	Shop & petrol pumps (community co-operative)	
District nurse	District nurse (based Kingussie)	
Doctor	Doctor & Associate	
Village Hall (needing repair)	New Community Hall	
1 pub	2 pubs	
1 hotel	2 hotels	Bed & Breakfast & Guest Houses
4 self-catering cottages	8 self-catering cottages	
2 holiday houses	4 holiday houses	1 caravan site
11 forestry houses	Forestry houses 3 privately-owned, 5 rented, 3 empty	All 11 now in use
2 empty cottages	Cottage now bothies	Bunkhouse for climbers
	5 new houses	

Table 1.4   Changes in Laggan, 1970–2000

Figure 1.75   BBC series *Monarch of the Glen*.

Forestry in the Cairngorms

Although forests in Scotland now cover 1.4 million hectares or 15% of the land area, this is still well below the extent of woodland cover of most countries in mainland Europe. In Scotland, the expansion of forestry during the late twentieth century was heavily criticised for the poor appearance of new planting, the intensive management techniques used, and the use of financial and tax incentives which encouraged low yield planting on poor quality land to the detriment of the natural environment.

The government's forestry policy is now based on multiple objectives: wood production, **biodiversity**, **landscape diversity** and recreational provision. Forestry supports over 10 000 jobs in Scotland, the majority of them in rural areas. Despite the use of mechanised techniques, the increase in timber production from 3 million cubic metres in 1994 to 8 million in 2014 will provide opportunities for further employment in wood harvesting and processing. A major problem now in many rural areas, notably Dumfries and Galloway, as timber harvesting increases is the poor condition of many minor roads and weak bridges which will have to carry the weight of timber lorries.

The concern for maintaining biodiversity and landscape diversity is partly a response to public concern about the huge swathes of monotonous coniferous plantations that appeared throughout the Scottish uplands during the post-war period, and also a recognition of the importance of sustaining and expanding native woodland habitats. Highland Birchwoods, a forestry company manages over 2000 hectares of native birch woodland. Recent examples of their successful projects include:

- the commercial cultivation of culinary shiitake mushrooms on small roundwood which would otherwise have had little value
- the use of new manufacturing techniques to produce flooring from timber which would otherwise have been used for firewood.

The importance of the woodland environment in providing opportunities for a wide range of recreational pursuits has long been recognised, with five Scottish Forest Parks, one of which is at Glenmore in the Cairngorms (Figure 1.71). The attraction of the Cairngorms for visitors is as much based on the fringe of varied woodland on the lower slopes of the hills as it is on the high tops. In terms of visitor numbers, with the exception of downhill skiing, far more people enjoy their recreation in the woodlands than on the hilltops. Glenmore Forest Park provides space for many activities, e.g. orienteering, that largely depend on the woodland environment. Development pressures are less than on the hilltops in that the woodland is able to absorb large numbers of people and their cars in sheltered and secluded surroundings have minimum impact on the natural heritage. In addition, the woodland is a rich reservoir for wildlife and also provides a bad weather alternative to the hills. Forest Enterprise has recently embarked on a strategy of enhancing the forest for amenity, conservation and recreation purposes in an attempt to reestablish a native pinewood. Other schemes include the introduction of the **Community Woodland Supplement** to the Woodland Grant Scheme. By these means the government encourages new woodlands which local communities can use for recreation. Over 1500 hectares of community woodland have been planted in Scotland, close to towns and villages since the scheme began in 1992.

Questions

12 Identify the similarities and differences in the problems facing Highland villages such as Laggan, and those in the Yorkshire Dales (pp28–29).

13 Why is UK forestry policy now based on multiple objectives, only one of which is timber production?

14 Referring to specific examples, identify successful attempts at forest diversification.

The Flow Country

Conservation issues in the glaciated uplands of the Cairngorms have received considerable media attention, but land use issues are no less important in other parts of Scotland where the threat to natural habitats has been just as great.

- **Dumfries and Galloway**
  14% of broadleaved woodlands felled and replaced by conifers; 39% of open heathland planted with conifers.
- **Caithness and Sutherland**
  More than 80 000 ha of the Flow Country has been planted with conifer trees.
- **Stirling District**
  40% of hedgerows in Stirling district destroyed for agriculture since 1970.
- **Grampian**
  20% broadleaved woodlands cleared for agriculture; 41% of hedgerows and treelines between fields destroyed.
- **The Lowlands**
  90% of grassland lost; 40% of heath destroyed; 50% of native woodland felled; 50% of peatlands destroyed.
- **Borders**
  21% of open heathland converted to poor grassland because of overgrazing by sheep.

The **Flow Country** of central Sutherland and Caithness is one of the most remote areas of the UK. It lies between the high glaciated mountains of north west Sutherland and the fertile farmlands of Caithness, forming a rolling plain of **blanket bog** (Figure 1.76), the largest in Europe (400 000 hectares). Glaciation here has left a very different landscape to that found in the Cairngorms and

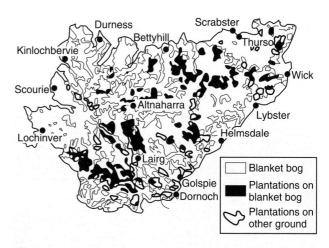

Figure 1.76 The Peatlands of Caithness and Sutherland.

further west. In this region the ice was warm-based and erosion took place over the entire landscape, resulting in **areal scouring**, although to no great depth and with relatively little change to the preglacial landscape. The area can be likened to a giant sponge in which the flat and wet peatlands are dotted with lochs, lochans and pools (Figure 1.77). This waterlogged landscape, whilst virtually useless for commercial farming, provides a unique habitat for breeding birds and is of international significance. Careful traditional management of the bogs for sheep, red deer and grouse has been possible without damaging the fragile ecology. The period since 1980 has brought about the most dramatic changes to this ancient scenery for 4000 years, principally due to the expansion of forestry.

Between 1979 and 1987 the Forestry Commission and a private forestry company, Fountain Forestry, bought up large tracts of land in Caithness and Sutherland and started draining and planting them. By 1987, one-fifth of this area had been lost to afforestation schemes. Until the late 1980s commercial forestry companies purchased large tracts of bog land at very low prices. Under the government grant system then operating for new forestry plantations, they were able to claim substantial tax concessions. The Forestry Commission planted 16 000 hectares and Fountain Forestry planted a similar area. The impact of forestry on the peatlands is severe and destroys the unique flow ecosystem. The deep ploughing and drainage, which precede planting damage the vegetation, lower the water table and bring to a halt thousands of years of peat formation.

Figure 1.77 The Flow Country.

Other effects spread beyond the plantations themselves and include river siltation and the **acidification** of freshwater habitats through fertiliser seepage. Research using pollen grain analysis from the peat bogs has revealed that woodland was never widespread in the area. The few trees which did grow died out about 4000 years ago following a change in climate resulting in colder and wetter conditions; this eventually led to the development of the peat bogs. Any natural woodland was restricted to the valleys and along watercourses. Some of the recent plantations have suffered heavily from attack by insect pests and from exposure to the wind. Much of the Flow Country is in the highest wind damage category area, meaning that many trees are likely to be blown down before they reach economic maturity.

On the positive side, the development of forestry brought much needed rural employment to an area with few job opportunities. By the mid 1980s Fountain Forestry alone had a workforce of 150. There was considerable media coverage about the threats to the peatlands which forestry was causing, and in 1988 the government withdrew the special tax

benefits for planting. This led to the virtual elimination of speculative tree planting in Scotland. As a result staff employed by Fountain Forestry fell to 23 by the early 1990s. New guidelines published in 1999 by the Forestry Commission on forestry and peatland habitats, taking account of recent EU and UK legislation effectively ban further woodland proposals in the area. Forest Enterprise have now begun to carry out selective tree removal and drain blocking within some of their plantations to help restore peatland damaged by forestry operations in the past.

In 1988, the Secretary of State for Scotland gave approval for a maximum of 175 000 hectares of peatland in Caithness and Sutherland (about half of the area of the unforested peatlands) to be designated an SSSI. By 1996 SNH had identified a total of 39 sites which have also been granted additional protection for breeding birds. The international significance of the peatlands has been further recognised through classification as a Wetland of International Importance and shortlisted as a World Heritage Site.

In 1992 SNH introduced The **Peatland Management Scheme** which operates in a similar way to Environmentally Sensitive Areas and aims to ensure a sustainable future for the Peatlands. Crofters, landowners and tenants voluntarily enter into a five-year agreement with SNH to manage their land in a traditional way, in keeping with its heritage value, covering grazing, peat cutting, moor burning and use of all-terrain vehicles. In return, SNH offers an annual payment of between £200–4000 depending on the size of the holding. As of August 2000, 123 such agreements were in operation covering almost 100 000 hectares of peatlands at an annual cost of over £150 000.

Although these new initiatives go a long way towards safeguarding the area from further afforestation, a damaging clash between developers and conservationists may now come from peat cutting for fuel. Small scale domestic peat cutting has been a tradition in the area for centuries, but is now noticeably declining as central heating offers a less labour intensive means of heating the crofters' homes. In contrast however, there is increased demand for industrially-cut peat from this area. One company which operates peat workings in the area is already exporting 10 000 tonnes of fuel peat a year to Sweden through the port of Scrabster, and wants to increase the size of its site to 900 hectares (Figure 1.78). The market for fuel peat is generally restricted by both high transport costs and lack of demand. However, with its lower sulphur content than either coal or oil, peat markets could be developed in the future using this 'green advantage'.

Benefits to the local economy through the expansion of peat cutting have to be balanced against the income derived from the development of high quality '**green tourism**' initiatives in the peatlands. The remote and wild qualities of the land, and the characteristic flora and fauna, in particular peatland birds, offer a unique experience for wildlife tourism. Any future development must be sensitively managed to avoid negative impacts on this natural heritage.

Figure 1.78   Industrial peat cutting in Caithness and Sutherland.

Figure 1.79   Forsinard RSPB reserve.

One successful recent project was the purchase of a 7100 hectares reserve at Forsinard by the Royal Society for the Protection of Birds (RSPB) in 1995 (Figure 1.79). Assisted by funding from the EU, a former railway station building has been converted into a visitor centre with a nature trail and regular guided walks in summer. In 1997 the reserve attracted over 4200 visitors and was estimated to contribute an additional £185 000 to the local economy.

The development of such projects is important in achieving the sustainable management of our natural resources and heritage. Almost all human actions affect the environment. The challenge is to assess the impact of such actions and to decide whether they are acceptable in the long term.

Questions

15 In what ways did glaciation leave a different landscape in Caithness and Sutherland to that in the Cairngorms? Give reasons for your answer.

16 Outline the physical and financial advantages which the Flow Country offered for forestry development during the 1980s.

17 Describe the impact of afforestation on
  a the landscape
  b the local economy of the Flow Country.

18 Why is the threat of further afforestation in the area now much reduced?

19 Outline the similarities between the Peatland Management Scheme and management of Environmentally Sensitive Areas.

20 Give examples of the development of Wildlife Tourism in the peatlands area and discuss whether this meets the criteria for 'sustainable development'.

Key terms and concepts

acidification	page 55
Agenda 2000	page 17
areal scouring	page 54
Area of Outstanding Natural Beauty (AONB)	page 8
bedding planes and joints	page 6
biodiversity	page 53
blanket bog	page 54
braided river channel	page 45
carboniferous limestone	page 21
chalk	page 6
clay vale	page 7
clearances	page 51
clint	page 23
Common Agricultural Policy (CAP)	page 17
Community Woodland Supplement	page 53
coombe	page 13
corrie	page 42
Countryside Agency	page 20
Cretaceous	page 12
dale	page 21
deposition	page 3
differential erosion	page 13
differential permeability	page 8
dip slope	page 6
dissected plateau	page 6
doline, or shake hole	page 24
downland	page 12
dry valley	page 13
English Nature	page 24
Environmentally Sensitive Area (ESA)	page 12
erosion	page 3
escarpment	page 6
farm diversification	page 17
Farm Woodland Scheme	page 18
fiord	page 40
Flow Country	page 54
fluvioglacial erosion	page 39
footloose industry	page 29
glacial drift	page 23
glacier	page 39
glaciokarst	page 21
global warming	page 39
gorge	page 23
Great Scar Limestone	page 21
green tourism	page 56
gritstone	page 30

S **ummary**

Having worked through this chapter, you should now know:

- the characteristics of scarp and vale (Jurassic limestone and chalk) landscapes, carboniferous limestone landscapes and glaciated uplands, and the formation of their associated physical features

- the factors involved in the evolution of the above landscapes; structure, rock type, geology, major geomorphological processes, climate, drainage patterns, soils and human factors related to changing land use and settlement

- the land resource base provides inter-related economic and social opportunities; agriculture (including the impact of UK and EU policies), forestry, recreation and tourism, mineral exploitation, industry and water storage

- competing demands for land result in environmental problems and conflicts within these landscape types. There is a need for protected areas such as National Parks

- UK and EU policies such as set-aside and ESAs affect rural land use.

Rural Land Degradation

Links to the core

This chapter deals with the causes and consequences of degradation of rural land, and looks at the development of conservation and management techniques to prevent rural land degradation. Since the causes of degradation may be due to physical processes or human activity, it combines many of the ideas developed in both the physical and human geography core. It builds on the concepts from the atmosphere, hydrosphere and biosphere units and integrates aspects of population geography and rural geography into detailed case study material from North America and Africa.

2.1 Land degradation – What is it?

Introduction

In a geographical context rural land degradation is linked to soil deterioration. Soil quality is said to be deteriorated if it cannot be used in the way it was previously used, or it cannot support the vegetation cover it previously supported. The processes which cause this deterioration may be natural events, e.g. floods, volcanic eruptions, or the result of human activity, or a combination of both. Whatever the cause, we need to understand:

- what factors are responsible for the degradation of the soil/land
- the mechanics of the processes involved in soil erosion by water and wind
- the factors that determine how susceptible a soil is to erosion.

Land degradation has an enormous social and economic impact on the people living in affected areas. This chapter will focus on three case studies:

- the Tennessee Valley, USA
- the Great Plains of the USA
- the Sahel region, North Africa.

Three themes will be followed in each case study:

The causes Why has land degradation occurred here? What physical and human factors are responsible? How do these factors work together to cause degradation?

The consequences What impact has degradation had on the landscape and on the people? What have been the social and economic consequences of land degradation?

The cures What has been done to prevent further degradation? What soil conservation and land management strategies have been put in place?

Climate plays a major role in forming soil (Figure 2.1). The process can take hundreds of years before a mature soil is formed. Like any material on the Earths surface, soils are exposed to the agents of erosion. If **erosion** happens faster than soil formation

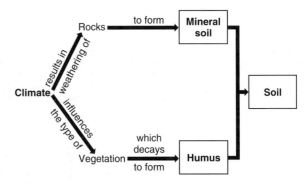

Figure 2.1 Soil formation.

then the soil begins to degrade. The quality of the soil will diminish, and, in certain circumstances, the cover of topsoil may be removed altogether, exposing the subsoil or parent rock.

Land degradation is the result. It is most common in the tropics and sub tropics, particularly in semi desert environments. It can have far reaching social and economic consequences for the people who live in affected areas, particularly in the developing world. Population growth is at its most rapid in the tropical and sub tropical countries. More people require more food, and so the land is used more intensively. This is the main reason for the rapid soil degradation taking place in these areas (Figure 2.2).

Degradation is damaging agricultural land very quickly (Figure 2.3.) Soils are under intense pressure as more food has to be produced on less land. Many farmers in the developing world are caught in a vicious downward spiral (Figure 2.4) which ultimately leads to degraded soils, low yields, low income, poverty and disease. Ironically, those areas which lose most soil are those which can least afford to do so. In the USA, 50% of fertiliser applied is used to compensate for loss in soil quality due to degradation. In Zimbabwe, Africa, loss of nutrients is three times greater than fertiliser input.

Such problems have led to the development of soil conservation and land management strategies ranging from capital intensive, high technology investments to low technology, labour intensive, small scale, local self help approaches.

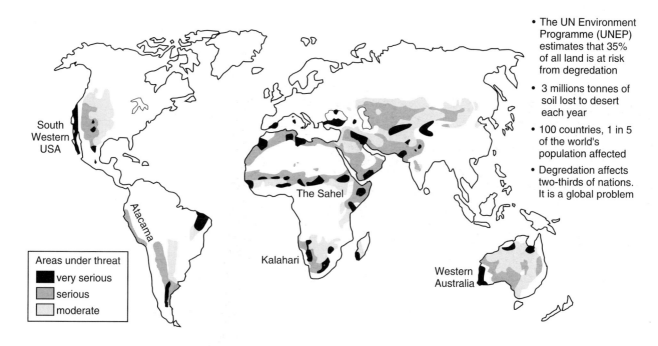

• The UN Environment Programme (UNEP) estimates that 35% of all land is at risk from degredation

• 3 millions tonnes of soil lost to desert each year

• 100 countries, 1 in 5 of the world's population affected

• Degradation affects two-thirds of nations. It is a global problem

Figure 2.2   The global distribution of areas of degradation.

The processes involved in soil erosion

A healthy soil is a living mixture of materials; inorganic particles such as sand or clay, decayed organic material (humus), air, water, micro-organisms, fungi and insects. All life in the soil is essential to maintaining fertility. Organisms mix nutrients through the soil, help ventilate it, add nutrients through animal waste and keep the structure crumby, i.e. capable of holding moisture.

Damage to soil life disturbs this ecosystem, and can lead to degradation and erosion.

There are three factors responsible for soil erosion:

● physical action
● chemical action
● biological action.

The impact of human activity on the land intensifies these actions, contributing to accelerated land degradation. Table 2.1 shows the human activities which have led to soil degradation across the world. Soil degradation in the developing world is a result of human activities such as deforestation and overgrazing (both brought on by population pressure). In the developed world, agricultural mismanagement is a much more important factor in land degradation. It is predominantly human activities which set in motion the various physical processes, e.g. erosion by wind or running water, which in turn cause the soil to be degraded or stripped from the land.

Biological action is usually the result of removal of vegetation cover. Vegetation cover protects the soil in many ways, for example, roots help bind the soil, foliage intercepts rain thus reducing raindrop impact, and the presence of vegetation protects the soil from wind.

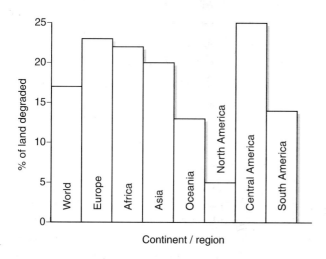

Figure 2.3   Human induced soil degradation, 1945–1990.

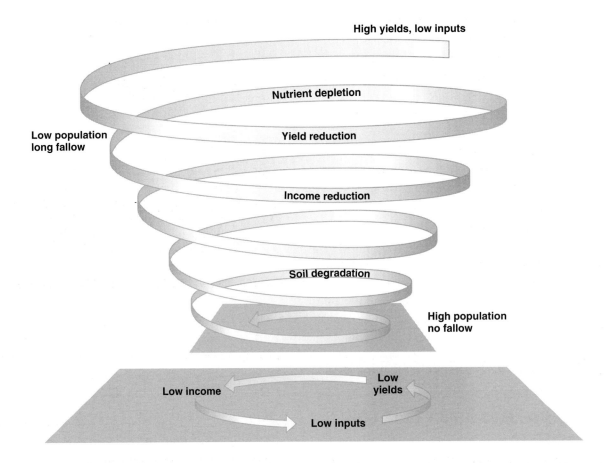

Figure 2.4 The downward spiral of poverty.

Removal of vegetation by humans or by natural processes expose the soil to wind and rain and increase the likelihood of soil erosion.

Farming activities can lead to both **physical** and **chemical erosion**. Persistent trampling by animals or humans and regular use of heavy machinery compact the soil and damage its structure. Root growth is held back and water is unable to drain freely through the soil, resulting in increased overland flow and more risk of erosion by running water.

	Deforestation	Overgrazing	Agricultural mismanagement	Overexploitation	Bio-industrial activities
Africa	67	243	121	63	+
Asia	298	197	204	46	1
South America	100	68	64	12	−
North and Central America	18	38	91	11	+
Europe	84	50	64	1	21
Australasia	12	83	8	−	+
World	**579**	**679**	**552**	**133**	**23**

SOURCE: OLDEMAN ET AL. (1990)

Table 2.1 Causes of soil degradation (in million hectares of affected areas)

Trampling and machinery destroy plant cover leaving bare patches exposed to wind and rain. Monoculture and over-cultivation weaken soils by removing too many nutrients. The soil becomes less able to support plant life and its healthy crumb structure breaks down leaving it at risk to attack by wind and water. Modern farming adds chemicals to the soil by way of pesticides, fertilisers and herbicides. These contaminate and damage soil life, i.e. bacteria, worms and insects resulting in a vicious circle of soil depletion. To maintain crop yields, more chemicals have to be applied and the vicious circle continues. Acid rain degrades soil quality, particularly if the soil is already acidic. Irrigation is often used to increase crop production, but if not properly managed, can cause **salinisation** of the soil. Unwanted salts rise to the topsoil, rendering it useless for agriculture.

Questions

1 a What is land degradation?

b Study Figure 2.2. Describe the pattern of land degradation across the continents of the world.

2 Complete the table below. Under each heading add examples of actions which can result in soil erosion.

Biological actions	Physical actions	Chemical actions

It is estimated that, in the last 50 years, human activities coupled with physical processes have degraded soils on 43% of the Earths vegetated land. The chief factors involved are:

- indiscriminate felling of rain forests
- **overgrazing** by livestock in semi desert environments
- inappropriate agricultural practices in all parts of the world.

Eroded soil must go somewhere. Damage is not restricted to the site of erosion, indeed, the environmental and economic damage suffered in places where the eroded soil is deposited may be just as severe.

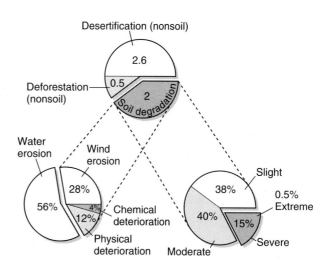

Figure 2.5 Global land degradation (billions of hectares).

Soil particles blown or washed away can clog up rivers, lakes and reservoirs or smother neighbouring landscapes in a blanket of fine dust. Figure 2.5 shows the causes of soil degradation across the world and the level of severity of the problem.

Approximately 85% of soil degradation is a result of erosion by water and wind and there are four different levels of degradation:

- slight degradation – most soils could be fully restored easily
- moderate degradation – considerable financial investment is needed to restore soil quality
- severe degradation – currently soil is useless for agriculture
- extreme degradation – incapable of restoration.

Soil erosion by water

There are four main types of soil erosion by water:

- **Rainsplash** is concerned with the impact of raindrops on the surface of a soil.
- **Sheet wash** is the removal of a thin, almost unseen, layer of surface soil.
- **Rill erosion** is the creation of very small eroded channels across a soil surface.
- **Gully erosion** is the creation of large gullies by large quantities of water flowing over the soil surface.

Soil erosion by water can be summarised as a three step process as shown in Figure 2.6.

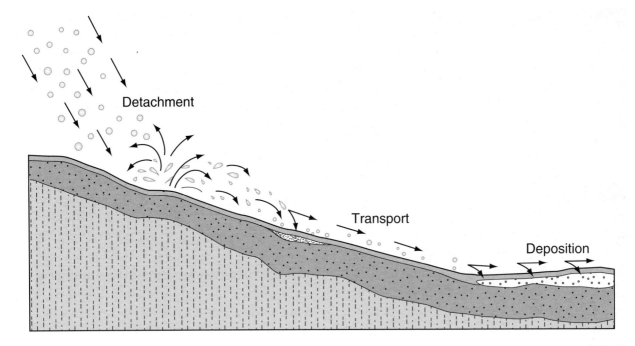

Figure 2.6   The three step process of soil erosion.

Step 1 Detachment Soil particles are detached from the main body of soil mass. This can be done by raindrops hitting the soil or by overland flow of water across the soil surface.

Step 2 Transportation Soil particles are carried downhill. They may float in water, roll or be dragged by water or splashed by raindrops.

Step 3 Deposition Eventually the soil particles are deposited in a downhill location. It may be on the bed of a river or lake, or on the sea-bed.

Rainsplash

The bigger the raindrop, the faster it falls and the harder it hits the soil. Large drops, such as those in a thunderstorm, hit the soil at about 30 km/hr with an explosive impact creating a mini crater. Soil particles are blasted outwards in a surrounding ring.

The force of rainsplash can be so great that:

• particles of soil become loosened from the main body of soil mass
• soil granules are broken down into finer particles – the soils 'crumb structure' is destroyed
• soil particles are exploded (splashed) at considerable distances from the point of impact, some as much as 2 m horizontally.

On a flat surface the impact of rainsplash is simply to move soil particles around in a random manner. However, if the soil is on a slope, then rainsplash will displace particles further downhill (Figure 2.7), resulting in the mass movement of soil downhill.

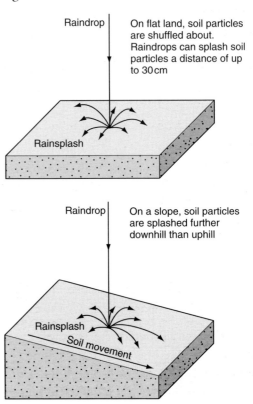

Raindrop | On flat land, soil particles are shuffled about. Raindrops can splash soil particles a distance of up to 30 cm

Rainsplash

Raindrop | On a slope, soil particles are splashed further downhill than uphill

Rainsplash
Soil movement

Figure 2.7   The effect of rainsplash on flat and sloping land.

Figure 2.8   Erosion caused by sheet wash.

Sheet wash

The rainsplash effect can also clog up the soil, leading to a form of soil erosion called sheet wash.
The tiny soil particles which are displaced during rainsplash find their way into the soil pore spaces located between the soil crumbs and clog them up. The result is the formation of a surface crust (perhaps only 1 mm thick), but the rate at which rainwater can now infiltrate the soil is reduced. The result is more surface runoff which picks up the displaced soil particles and transports them downslope. The fragile crust survives until the next rainstorm when it is disintegrated by the first few raindrops, only to be reformed again as rainsplash blasts a fresh set of tiny soil particles into the pore spaces.

If this happens on soils exposed to heavy rainfall where the slope is less than 5 degrees then sheet wash erosion will occur (Figure 2.9). Such slopes allow a very thin film of water to flow smoothly downslope (**sheet flow**). This process does not have the power to detach soil, but transports already detached particles.

Rill and gully erosion

These are the most serious forms of soil erosion by running water. Most land surfaces are irregular, with natural depressions, slopes and channels. A sudden burst of rainwater, which is unable to soak into the soil will flow over the surface, finding its way into natural channels. Here it will gather in volume, power and speed as it makes its way downslope. The flowing water detaches particles of soil and moves them downslope.

The result may be numerous small, eroded channels criss-crossing the landscape, no more than a few centimetres deep. These mini-channels are called rills. Rills are not permanent. Those formed in one rainstorm may be obliterated by the next storm, when an entirely fresh network of rills may be created, in positions unrelated to the previous network. They have no permanent connection to the drainage system. On farmland annual ploughing will cover up and obliterate rills that were formed over the previous season, but the damage has already been done, the soil has been eroded (Figure 2.10).

Gullies are steep sided water channels which carry water only during rainstorms and, unlike rills, are a permanent feature of the landscape. If the volume of runoff water becomes concentrated into one channel (on a slope of more than 5 degrees) the rushing water will cut deep into the soil. Gullies, several metres deep and wide are quickly carved out.

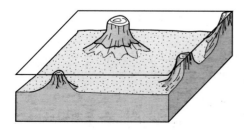

Figure 2.9   Sheet erosion.

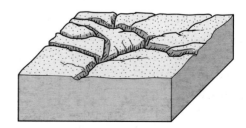

Figure 2.10   Rill erosion.

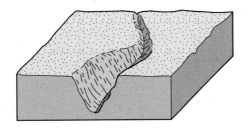

Figure 2.11a   Gully erosion.

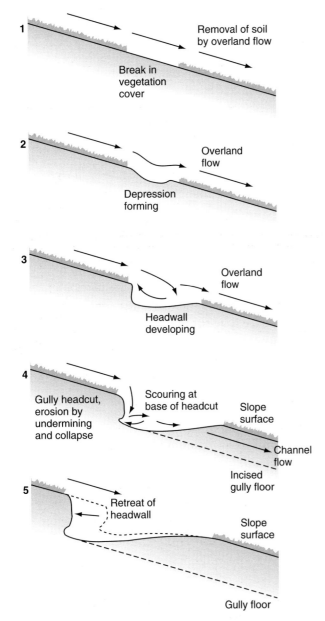

1   Removal of soil by overland flow

Break in vegetation cover

2   Overland flow

Depression forming

3   Overland flow

Headwall developing

4   Gully headcut, erosion by undermining and collapse

Scouring at base of headcut

Slope surface

Channel flow

Incised gully floor

5   Retreat of headwall

Slope surface

Gully floor

Figure 2.11b   The development of gullies.

Gullies can develop as enlarged rills. They can also be formed where a break in the vegetation cover on a slope allows runoff to carve out a mini depression. Over time this grows larger, allowing more rainfall to be trapped before making its way out of the depression to follow a channel (gully) downhill (Figure 2.11a). The gully can retreat uphill in much the same way as a waterfall retreats upstream. The headwall soil is progressively undercut, collapses into the channel and is washed downslope (Figure 2.11b).

Soil erosion by wind

Wind erosion is most common in arid or semi arid areas. The following conditions encourage soil erosion by wind:

- soil which is loose, dry and finely grained
- a relatively flat land surface
- little or no vegetation cover
- large fields
- strong winds.

Strong winds blowing over relatively dry soils pick up the finer particles and may transport them hundreds of kilometres across continents and oceans. The topsoil, the most fertile portion of the soil, is literally blown away. It is a worldwide problem often exacerbated by human misuse of fragile semi arid ecosystems, often brought about by population pressure.

As with water erosion, erosion by wind is a three step process:

- detachment
- transportation
- deposition.

The finest soil particles are transported in **suspension**. Wind picks them up, lifts them into the air (from a few metres, to several kilometres high), and may blow them horizontal distances of several hundred kilometres. Dust storms are a very spectacular visual indication of soil particles in suspension. These particles return to the ground only when the wind drops sufficiently or when rain washes them down.

The movement of larger particles by the wind by a series of short bounces or jumps close to the ground is called **saltation** (Figure 2.12).

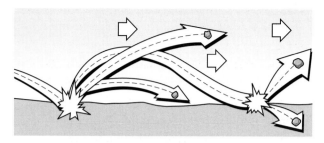

Figure 2.12   The process of saltation.

Saltation acts on particles which are too big to be carried in suspension in the air, but small enough to be picked up and raised off the ground by gusts of wind. They hit and move other particles as they travel.

The force of the wind may cause the largest particles to roll or slide along the soil surface in a process called **surface creep**. Smaller bouncing particles encourage surface creep when they hit the larger particles.

Although suspension is the most spectacular process of soil transportation it usually accounts for only about 15% of total movement, surface creep accounts for around 15% and saltation 70%.

Questions

3 a Describe the three-step process which is common to all types of soil erosion.

b With the aid of a diagram describe the rainsplash effect.

c Describe the conditions necessary for sheet wash to take place and explain the process.

d Complete the table below to compare rills and gullies. Add bullet points to bring out similarities and differences in their appearance and formation.

Rills	Gullies

4 Describe and explain the differences between these different forms of transportation of soil particles by wind:

- suspension
- saltation
- surface creep.

Factors affecting soil erosion

The extent to which any soil is likely to be eroded by water or wind is a combination of a number of physical and human factors (Table 2.2). The rate at which soil is eroded is a function of both **erodibility** (the resistance of soil to detachment and transport) and **erosivity** (the potential of slope processes to cause erosion). A light, dry, sandy soil with limited vegetation cover on a steep slope in an area of tropical downpours would have high erodibility and erosivity, whereas a fine grained clay on flat land with a dense vegetation cover in an area of gentle rainfall would have low erodibility and erosivity.

Unlikely	Soil erosion	Likely
LOW	rainfall intensity	HIGH
LOW	runoff volume	HIGH
LOW	wind strength	HIGH
HIGH	soil infiltration rate	LOW
DENSE	vegetation cover	NONE
GENTLE	slope angle	STEEP
LOW	population density	HIGH
GOOD	soil and land management	POOR

SOURCE: AFTER MORGAN (1991)

Table 2.2   Factors affecting soil erosion

Plant cover (vegetation)

The more dense the vegetation cover, the greater the protection the soil has from erosion. Just how much protection plants offer depends on the density of the ground cover, the density of the root systems and the height of the plants. Dense root systems tightly bind the soil particles together. A dense ground cover intercepts raindrops, absorbing their energy, thus protecting the soil underneath from the full force of impact. However, if the plant cover has a high canopy, e.g. in a deciduous forest of beech and oak trees, small raindrops collect on the leaves, join up to form large drops and fall 20–30 m to the ground, thereby creating a significant rainsplash effect.

Planting Density plants/m²	Corn	Soyabean	Oats	Peas
25	57%	65.9%	78.4%	78.9%

SOURCE: AFTER LAL (1988)

Table 2.3 Percentage of total rainfall penetrating a canopy of vegetation.

The most protective plant cover has a dense root system and a low canopy (under 1 m tall). Table 2.3 gives an indication of the percentage of rainfall which penetrates the canopies of various crops. Clearly, some crops offer better protection from rainsplash than others.

The relationship between plant cover and soil loss from water erosion is not a directly proportional relationship. Notice how, in Figure 2.13 even a small percentage of plant cover has a big impact on protecting the soil, a cover of only 10% gives 50% protection.

Vegetation also gives protection by protecting the soil surface from the effects of wind. On a bare, uneven soil (Figure 2.14) wind speed registers zero at ground level. However, the presence of plants significantly increases the zone of 'no wind' to about 70% of the height of the plant. Again, a small percentage of plant cover will give a relatively large protection factor.

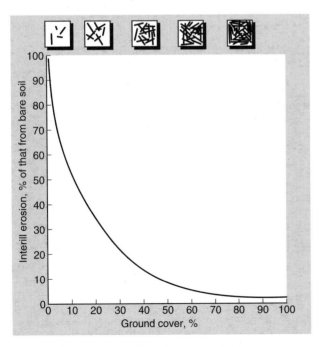

Figure 2.13 The effect of vegetation on soil loss.

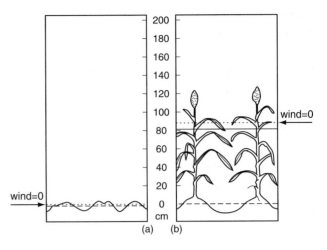

Figure 2.14 The effect of vegetation on wind velocity.

Rainfall and relief

The intensity of rainfall is an important factor influencing erosion. Intense rain has large drops which break up the soil crumb structure and there is more surface runoff to transport detached soil particles downslope. Gentle rain of low intensity causes little erosion, however torrential downpours may result in severe erosion. So much so that a handful of storms over a ten year period may account for most of the soil erosion in that time. Generally, the most erosive rains occur in tropical and sub tropical climates.

The steepness of a slope affects erosion much more than the length of the slope. Slopes most likely to encourage erosion are steep, short and convex. Erosion will be less on long, gentle, concave slopes and negligible on flat land. The impact of slope on wind erosion is complex but soils most at risk are those on flat open treeless landscapes where the wind has a long fetch.

Soil type and structure

The rate at which water drains down through the soil, i.e. its infiltration rate, is another important factor affecting erosion. A coarse sandy soil with large particles and pore spaces has a high infiltration rate, whereas a clay based soil with fine particles and small pore spaces has a low **infiltration rate** (Figure 2.15).

A high infiltration rate means less water available for **surface runoff**, and so less water erosion.

The structure of the soil is also important. Larger particles are less easily moved by water or wind than small particles. A coarse sandy soil is less likely to be blown or washed away than a fine grained cay or silty soil (assuming the soils are dry).

Questions

5 Construct a summary spider diagram which identifies all the factors affecting soil erosion.

6 Identify the ways in which vegetation cover protects soil.

7 Explain why:

a short bursts of intensive rainfall are very effective in eroding soil

b soils with a fast infiltration rate are less prone to erosion by water.

8 Study the table below giving details on three soils. Taking each soil in turn, assess the level of risk of soil erosion by water and explain your assessment.

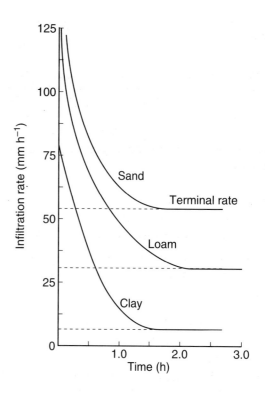

Figure 2.15 Soil infiltration rates.

Soil	Soil infiltration rate (cm per minute)	Vegetation cover	Rainfall
Soil 1	3.0	close cover shrubs, small trees, grasses	500 mm per year little seasonal variation low intensity
Soil 2	0.5	no continuous ground over, grass clumps	300 mm per year summer wet season high intensity
Soil 3	1.0	deciduous forest, sparse grass undergrowth	650 mm per year little seasonal variation low intensity

Case Study   Tennessee Valley

Location

The area of study is the valley of the Tennessee River in east central USA. The river is within the Mississippi River basin, draining much of central USA southwards into the Gulf of Mexico (Figure 2.16). The Tennessee River is extremely large. It rises on the western slopes of the Appalachian Mountains and flows westwards via a huge southerly loop, before joining the Ohio River, itself a tributary of the Mississippi. Although much of its 650 mile journey is within the state of Tennessee, a significant portion is in Alabama and its waters are also fed by streams originating in Georgia, North and South Carolina and Virginia.

Climate

Due to the sheer size of the river basin the climate can vary considerably. The climate graph for Knoxville (in the main valley floor in the upper part of the basin) is shown in Figure 2.17. On the higher ground in the Appalachian

foothills, rainfall figures would be higher (up to 2000 mm per year) and temperatures would be lower than for Knoxville. The climate graph for Glasgow is given for comparison, and the following points should be noted:

- The average temperatures are generally higher in Knoxville as it is much further south than Glasgow (Knoxville 36°N Glasgow 56°N).
- The temperature range is greater in Knoxville. This is because Knoxville, unlike Glasgow, is far from any cooling influence of the sea in summer and warming influence of the sea in winter.
- The rainfall totals are broadly similar. This disguises a very important feature of rainfall in the Tennessee Valley. Throughout the year, but especially in summer, rainfall comes in heavy thunderstorms as warm, moist, unstable tropical maritime air pushes northwards from the Gulf of Mexico. Short intense rainstorms on any unprotected soil on steep slopes will result in rapid soil erosion.

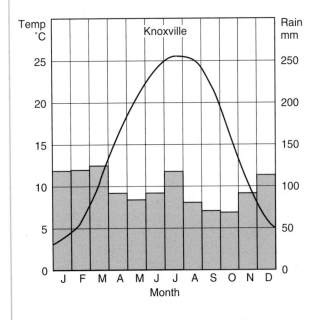
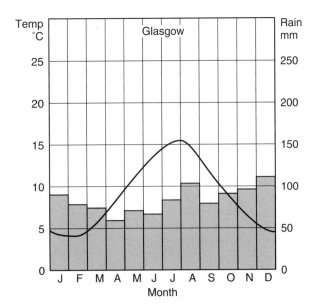
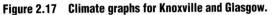

Figure 2.17   Climate graphs for Knoxville and Glasgow.

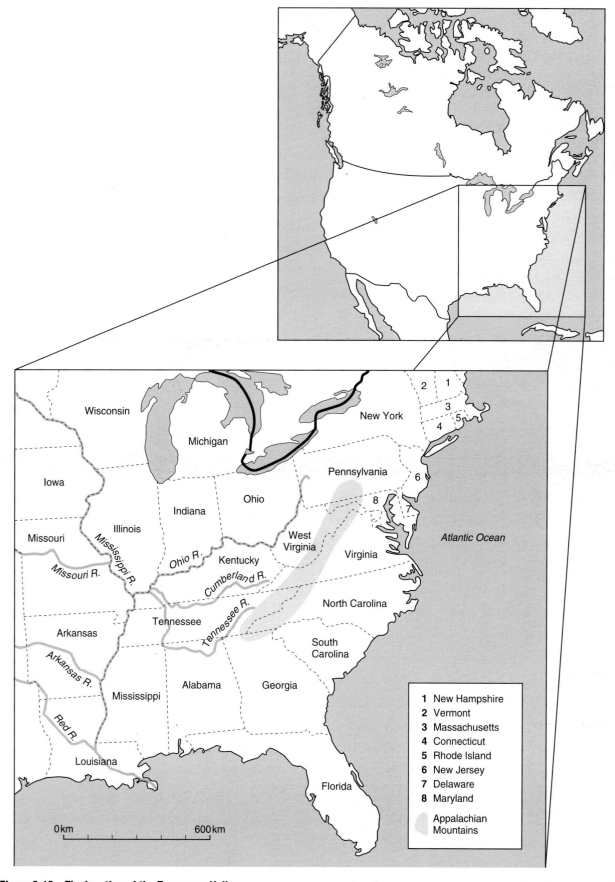

Figure 2.16   The location of the Tennessee Valley.

Soils and vegetation

The natural vegetation of the Tennessee region is forest; deciduous forest on the lower lying land and valley floors, which gives way to coniferous forest as altitude increases. This is similar to the vegetation pattern in the UK.

The pattern of soils is much more complex, but the main zonal soil is that associated with deciduous forest, namely the brown forest earth. In the areas dominated by coniferous pine forest, i.e. the high ridges, mountain tops and V-shaped valleys, the steepness of the relief prevents the formation of a true podzol (the soil associated with coniferous forest).

In the main valley floor, successive flooding by the Tennessee River has left a thick carpet of alluvial deposits spread over the flood plain (Figure 2.18).

Questions

1 Compare and contrast the climate of Scotland (Glasgow climate graph) with that of the upper Tennessee River valley (Knoxville).

2 Describe the differences in the nature of rainfall between these two areas. Why is the potential for soil erosion much greater in Tennessee?

Land use

To fully understand the causes of land degradation in the Tennessee Valley we need to look briefly at how the land was used until about 1930. There were three main land uses; farming, mining and forestry (Figure 2.19).

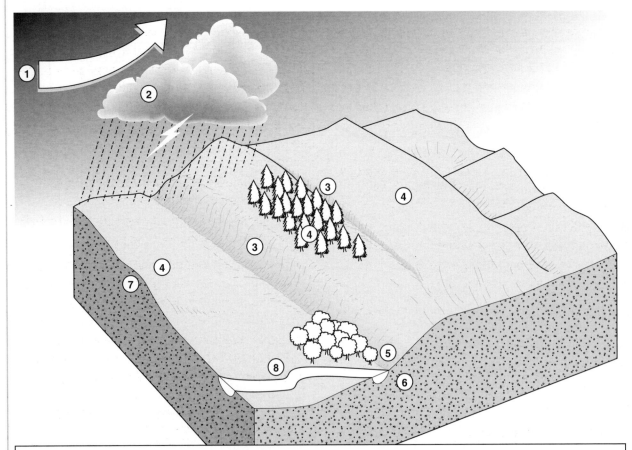

1 Warm moist unstable tropical maritime air from the Gulf of Mexico

2 Rain in intense thunderstorms, especially in summer

3 Steep sided V-shaped valleys

4 Coniferous pine forests dominate the steep slopes

5 Deciduous hardwood forest (oak and hickory) on the lower slopes and valley floor

6 Brown forest earth soils develop under the deciduous forest

7 Thin acidic soils develop under the coniferous forest

8 Alluvial soils on the floodplain

Figure 2.18 **The landscape of the upper Tennessee Valley.**

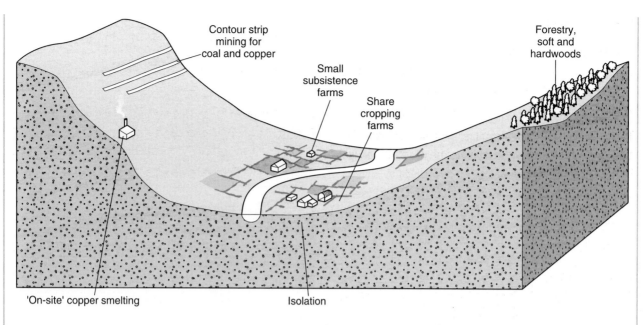

Figure 2.19   Occupations in the Tennessee Valley in the 1930s.

Farming Over 50% of the population lived on farms and, of these, 25% lived on farms they did not own. Farming was mostly self sufficient and confined to the lower slopes and valley floor (steeper slopes were forested). In some places soils were too poor to farm, while in others, e.g. in northern Alabama, the fertile soils supported cotton farming.

Mining Mining for coal took place in an area called the Cumberland Plateau, to the west of Knoxville. Coal seams were thin but close to the surface so mining was carried out in open cast strips which followed the contours of the landscape. This left ugly scars on the land. Mining for copper was carried out east of Chattanooga (at Copperhill on the Tennessee–Georgia border). Forests were cut to fuel copper smelting which was done 'on site'.

Forestry Forestry was a significant employer especially in the Blue Ridge area of North Carolina and in the Highland Rim area south west of Nashville. By 1930 lumber companies had logged most of the natural forest of soft and hardwoods in both of these areas leaving the soils, often on steep slopes, exposed to heavy rainstorms.

Social and eonomic conditions

The Tennessee Valley was a pocket of poverty and social deprivation in the 1930s. Incomes were well below the national average. Poverty was reflected in health; diseases such as TB, typhoid and malaria shortened life expectancy, malaria affected 30% of the population and malnutrition was widespread.

Birth rates were high (30% higher than the national average) as were infant mortality rates. Access to health care services was very difficult if not impossible for people on such low incomes. Education was poor; illiteracy rates were double the national average, and much of the valley basin remained an isolated 'backwater' essentially cut off from the rest of the country by a skeletal communication network. The river, which might have provided a communication spine was made useless for navigation by regular flooding along most of its length. The river regularly changed channel and in places became very wide and very shallow, less than 0.5 m deep.

These factors provided major disincentives to industrial development. As the USA industrialised in the early years of the twentieth century, Tennessee was ignored. Social and economic conditions became so harsh that some mountain villages were completely abandoned. Other villages and towns lost many of the younger population as they moved away to find work and a new life elsewhere in the USA.

The causes of land degradation in the Tennessee River basin

Land degradation in Tennessee was a combination of physical and human factors:

Physical causes The landscape of the upper Tennessee River basin is vulnerable to land degradation through soil erosion.

The reasons for its vulnerability are the numerous steep sided V-shaped valleys of the tributary streams and the nature of the rainfall, i.e. short heavy downpours. The role of vegetation is crucial here. The foliage and root systems of the dense coniferous forests in the upper basin protect the soil. Root systems help bind the thin soil together and foliage acts as a shield, protecting the soil from rainsplash impact during the heavy downpours. Unprotected soil is very vulnerable and quickly carried downslope by sheet wash, rill and gully erosion.

Human causes The human causes of land degradation are bad mining practices, bad forestry practices and, most significantly, bad farming practices.

Strip mining for coal left ugly open scars on the land. Trees were cleared leaving a bare land surface exposed to attack by rainwater. Mounds of mining waste suffered sheet and gully erosion, with silt being carried by fast flowing mountain streams into the main river itself. Copper ore was smelted on site with local wood being used as fuel. The newly exposed deforested soils were quickly washed from the hillsides. Poisonous fumes from the smelting process killed nearby vegetation, while poisonous sulphuric acids from the smelters were carried into the river.

Logging companies stripped the hillsides quickly and indiscriminately, so much so that by 1930 almost 50% of the original tree cover had been removed. Little attention was paid to proper forest management practices and replanting, which would have protected the soil and ensured a supply of timber in the years to come. The thin

Figure 2.20   Eroded land.

soils on the steep slopes no longer had the protection that vegetation cover offered, and heavy rains washed the soil into streams.

The worldwide economic recession of the 1930s saw mass unemployment in the factories in the industrial heartland of north eastern USA. As a result, former Tennessee emigrants drifted back to their homelands in the valley. New land on steep slopes was cleared to create new farms. Farming was barely above subsistence level, and farmers would abandon fields which had low yields, clear forest and try again, leaving the abandoned land to soil erosion. In total, about 12 mllion acres were cleared for farming.

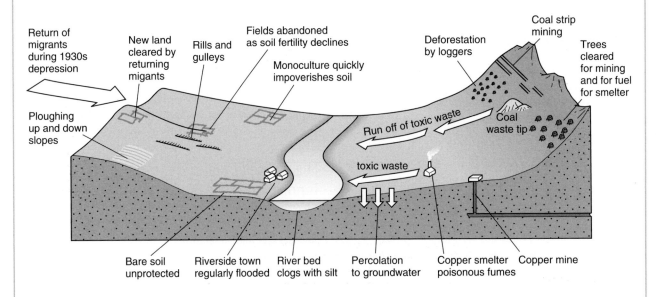

Figure 2.21   Causes of land degradation in the Tennessee Valley. C

Of these, 7 million needed erosion control measures and a further 1 million were eroded to the point of abandonment (Figure 2.20).

In addition, various bad practices employed by farmers contributed significantly to soil erosion:

- **Ploughing** Ploughing up and down the slope created furrows which became ready made channels transporting the rainwater downslope. In heavy storms the soils could not absorb the rainwater quickly enough so excess water ran downhill using the ready made furrows. Rills then gullies were quickly carved out as the soil was washed away.
- **Monoculture** The repeated planting of the same crop year after year resulted in a depletion of nutrients in the soil leading to a breakdown in soil structure, making it easier for wind and rain to carry soil particles away.
- **Fertilisers** At this time chemical fertilisers were not widely available and buying them was far beyond the means of most farmers. The application of available manure could not properly restore the soil. Soils became exhausted quickly and were very susceptible to erosion by wind and rain.
- **Soil exposure** The widespread practice of leaving the soil unprotected for part of the year led to sheet erosion becoming a common problem. This problem was particularly severe in the cotton growing lands of northern Alabama.

These causes of land degradation are summarised in Figure 2.21.

Questions

3 a What were the three main land uses in the Tennessee Valley in the 1930s?

b Make a bullet point list of the social and economic conditions experienced by people in the Tennessee Valley in the 1930s.

c Explain why industry was reluctant to move to the Tennessee Valley at this time.

4 What were the physical causes of land degradation?

5 'Bad practice' lay at the heart of the human causes of land degradation. Construct two spider diagrams, one to show bad farming practices, the other showing bad mining practices.

Consequences of land degradation

The result was massive soil erosion right across the Tennessee basin. Vast quantities of soil was stripped from the land and transported into fast flowing mountain streams. These streams channelled the soil and silt into the much slower flowing Tennessee River. Here it sank to the river bed, clogging up the channel forming numerous sand bars and mud banks (Figure 2.22). In places the river became extremely shallow and very wide as water spilled out of the clogged channel on to the floodplain. Flooding was a regular annual occurrence, sometimes bi-annual, across the entire valley downstream of Knoxville.

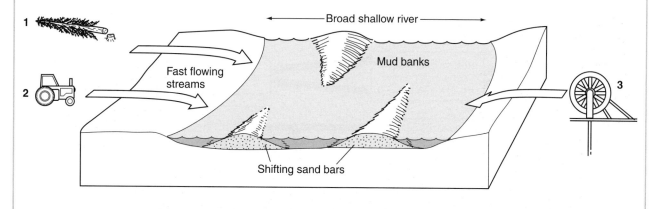

1	Soil and silt from bad forestry practice
2	Soil and silt from bad farming practice
3	Soil and silt from bad mining practice

Figure 2.22   River channel and sand bars.

Regular flooding had a devastating impact on the people and the economy in the Tennessee Valley. The river was unmanageable and its shallow depth and shifting sand bars made it useless for navigation. People living on the floodplain suffered most. This area contains the best quality farmland in the region, but each time it flooded farmers suffered ruined crops and lost harvests. This forced some farmers to move further up the valleys, clear new land and try to make a new start.

There was little incentive for investment in industrial development in the region because:

- communications were poor. Roads and bridges were often closed due to floods. The river, a potential source of cheap transport, was unnavigable. Railway companies were reluctant to invest in new track.
- the only power source available in the quantities required by industry was coal. It was bulky and expensive to transport, and waste had to be disposed of after burning.
- the people in the region suffered from poor educational standards and poor health. A large skilled work force would not be available without investment in education and training.

The root cause of all these problems was soil erosion. If this could be controlled, then the river could be managed, investment attracted and the social and economic recovery of the area begin. Figure 2.23 summarises the causes and consequences of land degradation.

Questions

6 a Explain how soil erosion in the upper Tennessee valley caused flooding downstream.

b Summarise the consequences of flooding for the following:

- farmers
- industry
- communications
- people.

Tackling erosion in the Tennessee Valley

An integrated plan was needed to solve the complex network of related problems in the Tennessee Valley, i.e. soil erosion, flooding, social deprivation, isolation, unemployment and a degraded landscape.

In 1933 the National Recovery Administration was set up with a budget of US $3 billion to spend on public works to get people back into employment. Part of this budget was spent on the creation of the Tennessee Valley Authority (TVA) in 1933. The TVA was originally given limited tasks:

TVA Act May 18th 1933

'To improve the navigability and to provide for the flood control of the Tennessee River; to provide for reforestation and the proper use of marginal lands in the Tennessee Valley; to provide for the agricultural and industrial development of the said valley'.

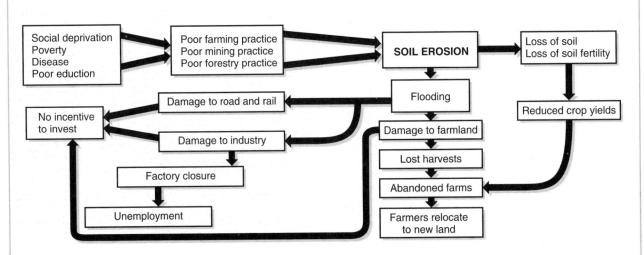

Figure 2.23　The causes and consequences of land degradation. ⓒ

However the TVA quickly and successfully expanded its remit to incorporate many other tasks:

- implement soil conservation strategies
- stem rural depopulation
- stimulate industrial regeneration
- improve the social and economic well being of the people.

This was the first US example of large scale government planning covering a complete river basin stretching across seven different states. The following text illustrates some of the schemes implemented by the TVA.

Reforestation

This was urgently required in the upper valley of the Tennessee River basin to stem the flow of eroded soil. Biological research suggested that the fast growing loblolly pine was the best tree to hold and trap soil. TVA tree nurseries were set up to provide a supply of trees, and millions of hectares were replanted on government lands over a 50 year period. Trees were provided free to private landowners so that they could plant shelter belts and reforest marginal hill land.

Education programmes were delivered to farmers providing information on good forest management practices. In addition to aiding soil conservation, the forests also improved water quality, regulated its flow, provided opportunities for recreation, created new wildlife habitats and formed a major scenic resource.

Reclaiming the mining landscape

Reclamation work on numerous abandoned mines was carried out by the TVA. This involved covering mined strips with soil and planting with pines – approximately 1000 trees per acre at an average cost of US $350 per acre (1965 prices). Since 1965 mining companies have had stringent reclamation requirements written into their contracts. The treatment of toxic metals and acids is paid for by applying a tax on mining companies proportional to the volume of minerals excavated. New mining techniques called 'cut and fill' which restore the landscape after mining are now widely practised.

Improving farming practices

Approximately 15 000 demonstration farms were set up across the TVA area. These served farmers by providing a local, easily accessible, open air classroom where the results of good farming practice could be seen and be evaluated. In addition the TVA ran an 'Agricultural Credit'

scheme. This allowed farmers to borrow money cheaply to improve their farms and invest in new technologies. Further assistance was offered through the Rapid Adjustment Farming Programme. This provided any farmer with free individual advice on which crops, livestock and land management techniques best suited the conditions on his land. 'Electro Development Farms' illustrated how the use of electrical equipment could save time and money while increasing productivity and living standards.

A number of soil conservation techniques were introduced on the demonstration farms, including:

- **contour ploughing** on the gentle to moderate slopes. This means that rainwater, which would otherwise run downslope causing the soil to be stripped away, is held in the furrows allowing time for it to soak into the soil.
- **strip cropping** on very gentle slopes and flat land. Tall crops are grown in narrow strips alternating with ground hugging crops. If water flows downslope the ground hugging crop will hold it back, giving it time to soak into the soil. The taller crop will also 'lift' any wind off the ground thereby reducing the likelihood of wind erosion (Figure 2.24).
- **intercropping** i.e. planting two crops together in the same field, but harvesting them at different times, ensuring the soil is covered at all times.

Figure 2.24   Contour strip cropping

- **double cropping** i.e. two crops are grown in the same field in different seasons thus ensuring continuous ground cover. Wheat and soya beans work well together.
- **crop rotation** helps maintain soil fertility and structure by the alternate withdrawal and replacement of different minerals by different crops.
- **chemical fertilisers** – government-run fertiliser stations conducted research and testing of fertilisers in order to produce 'tailor made' fertilisers to suit the needs of specific farms (Figure 2.25).
- **crop changes** i.e. farmers encouraged to change to crops which were better suited to the soils on their farm. Crops which quickly exhausted the soil (e.g. corn) went into decline while others like soya increased.
- **run off interception** i.e. particularly vulnerable land was protected by constructing concrete lined channels and ditches to intercept and control rapid **run off**.

Figure 2.25   Fertiliser demonstration plot.

Flood control, HEP and industrial development

The risk of flooding was reduced by soil conservation strategies in forestry, mining and farming. Flood control, river navigation and the generation of cheap hydro electric power (HEP) can only be achieved by building dams (Figure 2.26). A total of 30 dams were constructed on the Tennessee River and its tributaries. The majority of these were small dams in the upper valleys and their purpose was mainly to regulate river flow but also to generate cheap HEP. Larger downstream dams generated HEP, and also improved communications as roads were built over them and lock gates were constructed to allow river navigation.

Figure 2.26   Pickwick Landing Dam – an upstream dam on the Tennessee River.

Everything was now in place to attract large scale industrial development:

- the risk of flooding to factories and communication links was gone
- navigation was now possible from the Gulf of Mexico via the Mississippi, Ohio and Tennessee rivers all the way to Knoxville. This offered cheap transport especially suited to the movement of bulk raw materials or manufactured goods
- cheap HEP was available in vast quantities
- there was a ready supply of labour. Regular, well paid work was attractive to hard pressed farmers.

Figure 2.27   Installation of electricity lines.

Large scale manufacturing industry took advantage of these conditions and moved in. Aluminium smelting utilised local supplies of bauxite, cheap HEP and cheap transport as did iron and steel manufacturing, car assembly and the arms industry. In their wake came jobs though numerous supply and support factories and service industries.

Rural communities benefited from the availability of cheap HEP. The Rural Electrification Administration brought electricity to villages and farms for the first time (Figure 2.27). Farmers were able to invest in new technologies, helping them to improve productivity and increase efficiency. Electrified farms increased their income by up to 200% in five years, the biggest single benefit being the introduction of refridgeration. The amount of farmland was reduced, however, as land was flooded by the creation of new reservoirs. Farmers on the more marginal hill lands abandoned the land to take advantage of regular well paid work in the new factories in the lower valley.

Effectiveness of the TVA

The TVA successfully implemented an integrated plan to solve the complex network of inter-related problems in the region. It put in place a range of strategies which conserved soil, controlled flooding, stimulated agricultural and industrial growth, stemmed rural depopulation and greatly improved the social and economic well being of the people. As an exercise in large scale, government-led social, economic and environmental regeneration it is an outstanding success. Its work is not over. Recently, soil erosion problems resurfaced when some farmers were tempted to overwork the soil in pursuit of large profits from growing soya beans. Programmes are still in place to continue research and farmer education in soil conservation techniques, to monitor soil erosion and to reclaim damaged land. A spin off benefit from all of this work has been the development of recreational facilities. Forests, reservoirs and rivers provide the environment for numerous forms of outdoor activities, bringing more employment opportunities and income to the valley. The area now caters for five million visitors per year. Figure 2.28 summarises the main changes brought to the valley by the TVA.

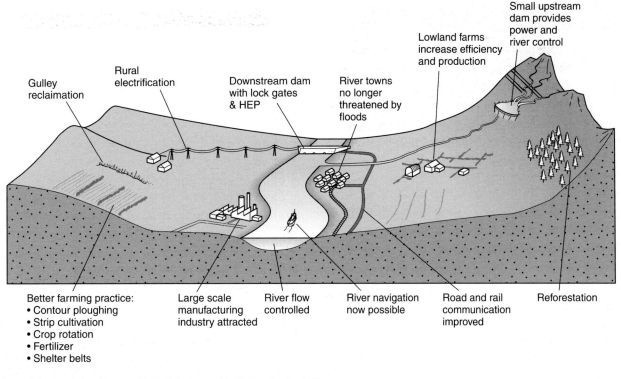

Figure 2.28   Changes brought to the Tennessee Valley by the TVA.

Q uestions

7 a What were the main elements of the TVA's integrated plan?

b Describe the steps that were taken to reforest the Tennessee Valley.

c Mining companies must now clean up their own mess. Describe how this is done.

d Draw a spider diagram to summarise the programmes and techniques used to improve farming practice.

e Explain why flood control was crucial in attracting industrial development to the Tennessee Valley.

8 How effective has the TVA been in achieving its aims? Explain your answer.

D ecision making

On 7–8 January 1998 heavy rain fell on the drainage basin of the Doe River in Tennessee (the Doe is a tributary of the Tennessee River). Figure 2.29 shows the hourly histogram of rainfall at Burbank in the river basin and Figure 2.30 shows the level of the Doe River over the same time period at Elizabethton, 10 miles downstream from Burbank.

● Comment on the relationship between the pattern of rainfall and the level of the river over the time period shown.

● Comment on, and suggest reasons for the length of the time lag.

● Construct an inter linked flow diagram to show how elements of the TVA scheme are connected. Use the example provided as a starter.

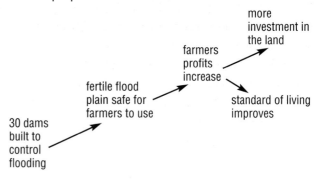

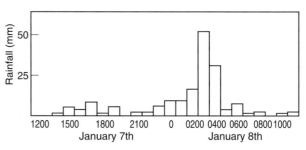

Figure 2.29   Rainfall in Burbank, Tennessee, January 1998.

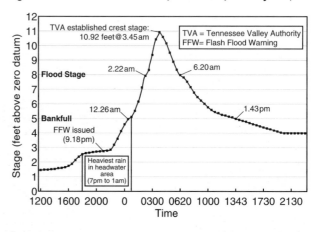

Figure 2.30   The level of the Doe River, 7–8 January 1998.

Case Study   The Great Plains

The Great Plains cover an area of approximately 36 million km² extending for 3870 km north to south and 1612 km west to east (Figure 2.31). With a population of approximately 10 million people, population density is extremely low. This region supplies around 25% of the world's wheat, oats, sorghum and corn. Rural land degradation in the Great Plains is a combination of physical, climatic and human factors. It is estimated that 1.7 million tonnes of topsoil are either blown or washed away in the USA each year.

Climate

Owing to the wide range of latitude, climate varies across the Great Plains, but in general terms it can be described as continental. As Figure 2.32 shows, annual precipitation totals are moderate (400–800 mm) while the annual range in temperature is high (−10°C to 30°C). Winters are long, dry and very cold. Owing to the low temperatures snow is common and can lie for up to five months. The summer months are hot with temperatures often exceeding 30°C.

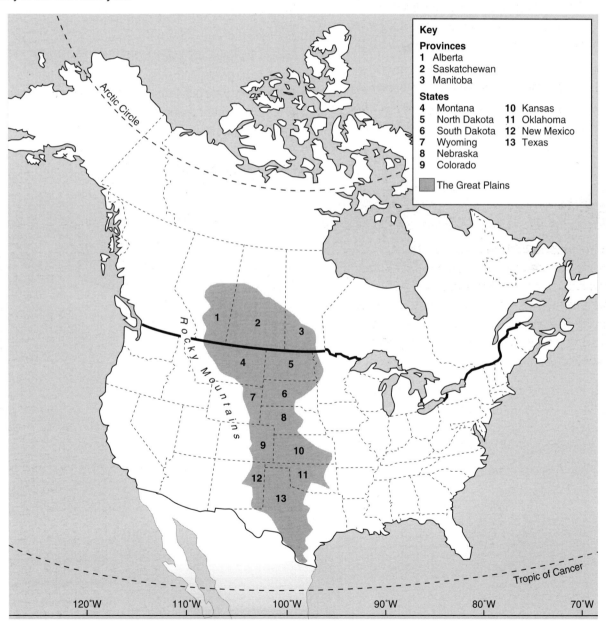

Figure 2.31   The location of the Great Plains of the USA and Canada.

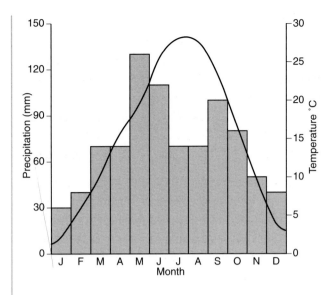

Figure 2.32   Climate graph – Oklahoma City.

In July and August precipitation is heavy but variable. Evaporation rates are high, often resulting in a moisture deficit but this is offset by snowmelt in early spring.

Natural environment

The Great Plains are predominantly an area of undulating grasslands. The density and range of species of the vegetation varies according to altitude, latitude and climate. The natural ecosystem of the Great Plains is diverse. The fertile soil supports a rich grassland cover as well as a wide variety of insect life and biota. The dominant herbivore was bison, which grazed the grasslands.

Chernozem is one of the main soil types found in the Great Plains (Figure 2.33). It is one of the most fertile soils in the world. It is black, has a deep humus layer and crumby texture which makes it both fertile and easily cultivable. The decomposition of the grass provides the rich humus layer.

Rainfall increases from west to east, and enables the region to be divided into three broad vegetarian bands:
- the short-grass prairie in the rain shadow of the Rockies,
- the mixed-grass prairie in the central Great Plains,
- the tall-grass prairie in the wetter areas to the east.

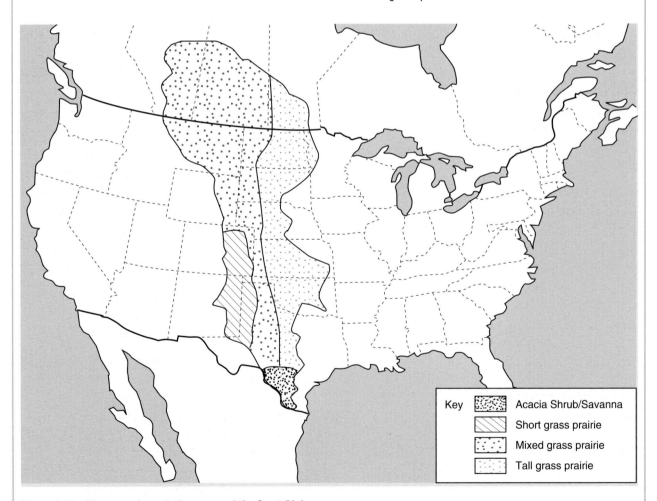

Key

Acacia Shrub/Savanna

Short grass prairie

Mixed grass prairie

Tall grass prairie

Figure 2.34   The natural vegetation zones of the Great Plains.

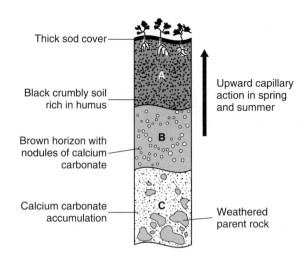

Thick sod cover

Black crumbly soil rich in humus

Brown horizon with nodules of calcium carbonate

Calcium carbonate accumulation

Upward capillary action in spring and summer

Weathered parent rock

Figure 2.33   A chernozem soil profile.

These three zones broadly correspond to the western Rangelands (largely livestock farming), the Wheat Belt and corn/soyabean areas respectively (Figure 2.34).

Human activity

The traditional hunter/gatherer lifestyle of the North American Indians was in harmony with the Great Plains ecosystem. It caused no long term damage to animal numbers, soil or vegetation. However, as North America was colonised, migrants moved west and the agricultural potential of the Great Plains was discovered. Settlers established cattle ranches, forcing the North American Indians off their homelands and slaughtering the majority of the bison herds.

By the late nineteenth century the settlers had realised the great potential in the Great Plains for the cultivation of wheat. The low lying relief, fertile soils and grassland vegetation led farmers to believe that the Prairies were ideally suited to arable farming. Initially the switch to farming was very successful but the farmers failed to take into account the variability of rainfall of the area. This, combined with several other factors as outlined below, led to severe land degradation, which North America is still dealing with.

The causes of rural land degradation in the Great Plains

Rural land degradation in the Great Plains cannot be attributed to any one factor but is a combination of physical, human and economic conditions.

Physical factors

Annual precipitation has varied greatly in the Great Plains between 1900–2000. The average annual precipitation during this period was 390 mm. During periods of below average rainfall e.g. 1930s, when average annual precipitation was less than 200 mm, soils dry out and lose adhesion. As a result the soil particles are more loosely bound and therefore more susceptible to **aeolian erosion**.

In contrast, during periods of significantly above average rainfall e.g. 1995–1999, when precipitation was above 600 mm, the land is subject to fluvial erosion as the intensity of the precipitation exceeds the infiltration rate of the soil. Colorado received 100 mm of rainfall on 4th August 1999 alone!

Human factors

The success of extensive commercial farming in the Prairies depends on growing wheat on a very large scale. However, the cultivation of the same crop on the same fields year after year (**monoculture**) depletes the soil of vital nutrients. As a result the soil structure is weakened, crop yields are reduced and the land degraded. In addition, the root network of the natural grassland is much denser than that of the wheat and so under cultivation, the soil becomes more loosely bound.

In preparation for planting, the fields are cleared and ploughed. This breaks up the crumby structure of the soil and leaves the fine, fertile topsoil particles open to wind erosion. Periods of low rainfall and high temperatures enhance the loss of soil as it dries out. Ploughing produces furrows in the fields. When it rains, the water is channelled into the furrows and runoff, carrying soil with it, is increased. This is a particular problem where the furrows are cut at right angles across the natural contours of the land.

As herd sizes increased in parts of the western Prairies, natural grasslands were overgrazed down to their roots, leaving the soil unprotected and open to wind erosion.

The growth of extensive, commercial agriculture was highly profitable, and the USA became one of the world's largest exporters of wheat. Between 1914–1917 wheat production in Russia was severely depleted due to World War I, and as a result there was a shortage of wheat on the global market thus the price rose sharply. This led to pressure being applied to North American farmers to increase productivity. This drive placed additional pressure on the land and resulted in land degradation.

1 What physical and climatic characteristics of the Great Plains make it an area which is susceptible to rural land degradation?

2 What lessons, if any, should the new settlers in the Great Plains have learned from the traditional way of life practiced by the native American Indians?

3 Were economic, climatic or agricultural factors the biggest contributor to land degradation?

Consequences of land degradation

Physical impacts

The most obvious and devastating impact of land degradation in North America has been the loss of topsoil. During the 1930s millions of tonnes of soil were blown across huge distances, smothering vegetation, choking wildlife, burying buildings and blocking out sunlight. The region became known as the **dustbowl** (Figure 2.35).

Figure 2.35　The dust bowl.

"Home was the safest place to be, but even that offered only limited sanctury. The dust was so thick that it filtered into houses as if walls were made of cheesecloth, not stone. The rags stuffed under doors and windows could no longer hold it out. Sand swirled into closets and cupboards, leaving its grimy fingerprints on dishes and clothes, sparing neither porcelain bowls lovingly passed through generations nor Easter dresses purchased yesterday."

"Art Leonard sat in the middle of his family's tire store, listening to the thrashing winds and gazing at the miniature dust storms billowing about the room. Those daring to peek through open doors were stung by sand on their faces and bare legs. Many would later talk about how as they peered out the windows, they saw Kansas, Oklahoma and little bit of Texas roll by."

The loss of soil fertility caused by wind erosion led to an increase in the use of fertilisers and pesticides, as farmers fought to sustain crop yields. In the long term this has resulted in the contamination of soils, rivers and aquifers. This in turn has affected indigenous flora and fauna, raising concerns about the environmental impact of extensive commercial agriculture.

Social and economic impacts

The large scale erosion and land degradation effectively wiped out the livelihoods of thousands of farmers and related services, e.g. shopkeepers and grain merchants across the Great Plains. They were left with the stark choice of abandoning the area in search of alternative employment or staying and attempting to rebuild their businesses. Many of those who left migrated to cities such as Detroit, in search of jobs in the car industry, or to Los Angeles. Farms were sold cheaply or even abandoned. This pattern of events led to rural depopulation across the region, and an increase in average farm sizes as farms were amalgamated.

The social costs of land degradation were high. Many people suffered from depression, suicides were common and unemployment rose sharply. The US Government was faced with the burden of providing financial support to those effected. Economically the USA was hit very hard, as it lost the valuable revenue normally brought in by the export of grain to regions such as Europe.

Solutions to the land degradation

Afforestation

One of the most effective ways to reduce the risk of land degradation is the planting of trees, and in particular **shelter belts**. Given the low-lying relief of the region and the strong winds which blow across it, the creation of shelter belts along field boundaries can reduce soil loss dramatically. Research has shown that trees are able to slow down wind speeds for a distance equivalent to ten times the height of the trees. However their effectiveness is dependent on the formation of the shelter belt, as shown in Figure 2.36. Although extremely effective and relatively cost efficient, the main disadvantages of shelter belts as a technique are the length of time required for the trees to mature, the maintenance they require and the cropland taken up by the trees themselves.

Contour ploughing

By ploughing along the lines of the contours instead of across them, farmers can reduce the amount of soil lost to overland flow by up to 50%. Contour ploughing reduces the speed of water flow across the land surface and therefore the erosivity and carrying capacity of the runoff is reduced. In addition to this, infiltration rates are greater, and growing conditions are enhanced by the increase in soil moisture. The only significant disadvantage of this technique is the additional workload involved in the identification, marking out and ploughing of the contours.

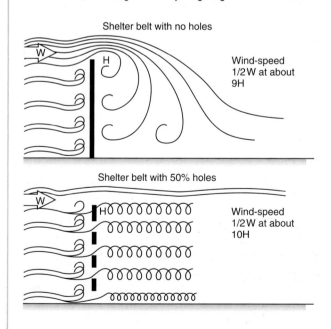

Figure 2.36   Different formations of shelter belt.

Figure 2.37   Laser levelling.

Laser levelling

Economically developed countries like the USA and Canada are in the privileged position of being able to employ 'high-tech' solutions to the issue of land degradation. Laser levelling is one such technique (Figure 2.37). The aim is to achieve as level a surface as possible in the fields. Lasers are used to control the height of rollers, which are towed across the surface of the soil. Where the laser detects an increase in the height of the soil surface the rollers are lowered and where a lowering of the surface is identified, the rollers are raised. The extremely flat surface produced means the effects of rainsplash and overland flow are minimised. However the equipment involved is expensive.

Minimum tillage

The principle of minimum tillage is based on limiting the number of trips made by machinery across the surface of fields in preparing, planting and harvesting crops. An example of minimum tillage is the **sweep plough**. Traditional ploughs work by turning the soil over in preparation for the planting of the following season's crop. This process breaks up the crumby structure of the chernozem soil leaving it exposed to processes such as rainsplash, overland flow and aeolian erosion. The sweep plough has been designed to cut below the surface of the soil, severing the roots of the previous crop and any weeds, without disturbing the soil surface. The fact that the stubble is left standing means it is able to afford some protection to the soil, as the new crop grows. Over time the stubble will decay, returning nutrients to the topsoil. The practice of leaving crop debris lying on the surface of the land is known as **mulching**.

Figure 2.38   Minimum tillage.

Arable farmers now have access to machinery designed to sow seeds, spread fertiliser and add pesticides simultaneously (Figure 2.38). This cuts down on the number of trips the farmer needs to make across the fields, minimising soil disturbance and compaction.

Fallow period

In the drive to maximise yields, many farmers in the Great Plains cut down or eliminated altogether a fallow period from their cycle of cultivation. This resulted in soils becoming exhausted, yields falling and land becoming degraded. Organisations like the **Soil Conservation Service** (SCS), set up by the US Government in the aftermath of the dust bowl, now encourage farmers to ensure land is left fallow on a rotational basis.

Intercropping and strip cultivation

This normally involves the cultivation of two or more crops in the same field (Figure 2.39). One such combination practiced in the Prairies is corn or soyabeans along with oats or legumes. The idea is to use crops which grow at different rates and/or are harvested at different times of year so that the field is never devoid of plant cover. Once again, this reduces the effects of wind and water erosion. In tests it has been found that in parts of the Great Plains

intercropping has resulted in yields of between 10–40% greater than from cultivation of one crop.

Cell grazing

This system, used by ranchers in areas such as Colorado, is designed to combat the effects of overgrazing. The rancher divides his land up into cells which are centred around a central corral (Figure 2.40). The cattle are restricted to one 'cell' at a time and allowed to graze that area for three days. They are then moved to another cell and the grass is given time to recover and rejuvenate.

Figure 2.39   The practice of intercropping.

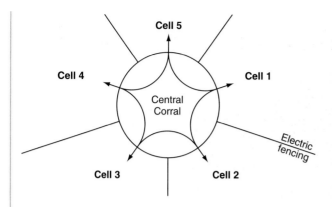

Figure 2.40   Cell grazing.

Figure 2.41   Boom irrigation in Colorado.

This system is relatively simple but very effective, providing the rancher has sufficient land to create the number of cells necessary to make it work and the resources available to fence off the cells.

Bison ranching

Ranchers in some areas of the USA, such as Texas are now rearing herds of Bison instead of cattle in a bid to prevent overgrazing. Bison are one of the indigenous species of the Great Plains and existed by not overgrazing the natural grasslands. What the ranchers have found is that Bison will graze and move on from one area to the next before its vegetation cover is irreplaceably damaged. The higher yielding strains of beef cattle brought into the area tended to strip the land of all cover (including the roots) thus leaving it susceptible to wind erosion. The main drawback to this technique is the low profitability and marketability of the bison meat in comparison to beef cattle.

Boom or centre-pivot irrigation

If an arable farmer is able to eliminate the potential risks of drought by irrigating crops then he/she is able to sustain crop cover and therefore minimise degradation from wind erosion. This can only be achieved in locations where a suitably large and sustainable source of water is available, such as a surface reservoir or underground aquifer. In areas such as northern Texas and New Mexico this has

been made possible by extracting water from the Ogallala Aquifer. As Figure 2.41 illustrates, boom irrigation has resulted in one of the most spectacular farming landscapes in the world. However, this technology comes at a cost. Not only is the equipment required to pump and distribute the water very expensive, but in many areas the rate at which water is being removed from the aquifers greatly exceeds the rate at which the water is replaced by precipitation. In time this can cause the groundwater to become contaminated and the water table to drop. Irrigation can also lead to the salinisation of soils as the irrigation water, which contains dissolved salts from the underlying geology, evaporates leaving the salts behind.

Education

Informing farmers of the long term consequences of their farming techniques as well as methods to minimise these consequences, remains one of the most effective ways of combating land degradation. This is an integral part of the work done by the Soil Conservation Service (SCS). Representatives from the SCS offer farmers advice and support in employing farming methods which will ensure the long term sustainability of agriculture in the Great Plains. This type of work is an on-going process. Memories can be short and if individual farmers have no experience of the damage and hardship which can result from inappropriate farming methods, they may 'cut corners' in order to increase profits.

The Sahel

In arid climates land degradation can lead to **desertification**. Fundamentally, desertification means land turning to desert. Desertification threatens approximately one third of the Earth's surface, affecting the lives of 1.2 billion people. As Figure 2.2 shows, the regions most susceptible to the process of desertification are those bordering the world's hot deserts e.g. the Sahel, which borders the Sahara.

It is estimated that 6 million hectares of productive land are lost each year to desertification. The Food and Agricultural Organisation of the United Nations estimated in 1998 that, if world soils continue to degrade at their present rate, in around 20 years the world will be unable to produce enough food to feed the rapidly increasing global population.

A combination of physical, climatic and human factors have made the Sahel region of Africa particularly susceptible to desertification.

Location

The Sahel runs parallel to the southern edge of the Sahara Desert from the west coast to the east coast of Africa. The average width of the Sahel is 500 km.

Its proximity to the Sahara is one of the principal reasons it is prone to desertification (Figure 2.42).

Climate

The principal features of the Sahel climate are shown in Figure 2.43.

- total annual rainfall is low, ranging between 200 mm in the northern Sahel to 500 mm in the southern Sahel
- rainfall is unreliable and highly variable
- rainfall often occurs as an intense tropical downpour, creating high potential **erosivity**
- a distinct dry season exists, i.e. rainfall only occurs during certain months. There are high rates of evaporation due to high temperatures which means much of the precipitation is lost.

The rainfall pattern in the Sahel region is a result of the **Intertropical Convergence Zone** (ITCZ) which migrates across the region each year. During the dry season, **Tropical Continental air** is dominant along with dry **Harmattan winds** blowing south across the Sahara.

Natural environment

Due to the variability of the rainfall across the Sahel, the soils and natural vegetation also vary. In southern regions where the rainfall is more reliable, well developed, fertile soils are found with a greater variety of flora and fauna.

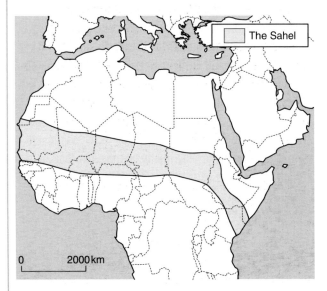

Figure 2.42   The location of the Sahel region.

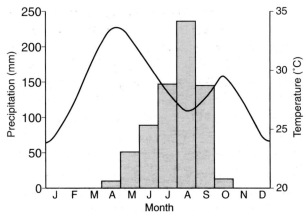

Figure 2.43   Climate graph for Sokoto in the Sahel.

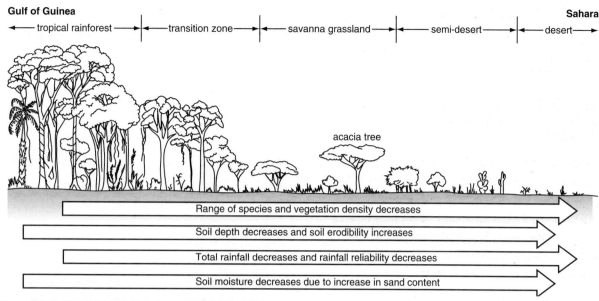

Figure 2.44   Natural vegetation zones in Western Africa.

Moving north, vegetation becomes less dense and the soils have a low organic and clay content (Figure 2.44). This means they are more susceptible to the processes of erosion.

Questions

1 What is desertification and why does it affect the Sahel?

2 **a** Using Figure 2.44, describe the features of climate in the Sahel that make it susceptible to desertification?

   b The climate graph does not tell us anything about rainfall reliability. Why is this information crucial in being able to understand why the Sahel is affected by desertification?

Human activity

Traditionally there have been two main types of subsistence agriculture practiced in the Sahel.

1. Arable farming

A wide variety of crops e.g. barley, sorghum and vegetables are grown together with drought resistant species e.g. millet. Trees are retained and/or planted amongst the crops. These provide protection from wind erosion by acting as a windbreak, binding the soil together and by shading the ground (which reduces evaporation). The trees will also provide a source of fuel for cooking and fruit/nuts for the farmer. Farmers leave sections of their land fallow for up to 20 years to allow it to rejuvenate.

It is important to grow a variety of crops in order to avoid the impact of **monoculture**, i.e. relying on only one crop, and exhausting the soil by removing the same nutrients year after year. The removal of these nutrients destroys the soil structure and leaves the soil more susceptible to wind and water erosion. The other advantage of growing a variety of crops is that if a crop is affected by disease or pests the farmer still has other sources of food for his/her family.

In this way the arable farmer is able to provide a mixed diet for his/her family while retaining the productivity of the soil by varying the crops grown.

2. Pastoral farming

Livestock rearing of cattle, goats and sheep is also practiced in the Sahel. The majority of these pastoral farmers are **nomadic**, moving from one area to the next following the rains, to find fresh pasture for their animals. The success of this type of farming depends on the following:

- avoiding overgrazing an area, leaving the soil vulnerable to erosion, as the vegetation no longer offers any protection
- finding sources of water for animals to drink e.g. wells, rivers, lakes
- the ITCZ providing sufficient rainfall for vegetation to grow in the areas traditionally used by nomadic pastoralists
- maintaining herd sizes at a level which is sustainable.

In many regions of the Sahel, the size (not the quality) of an individual's herd is a measure of his status within a tribe. It is a hard but highly adapted way of life. Some tribes such as the Samburu of northern Kenya are able to survive for long periods on what their cattle can provide e.g. they drink a mixture of the blood and milk, eat the meat and wash in the urine.

Q u e s t i o n s

3 Explain why the traditional agricultural techniques practised in the Sahel have been successful for many centuries and have not resulted in the long term damage of the Sahel's ecosystem.

Causes of rural land degradation in the Sahel

The causes of rural land degradation in the Sahel cannot be attributed to any one factor, but the interaction of a wide range of both physical and human variables.

Physical factors

Droughts are now occurring in the Sahel with greater frequency than has been experienced previously. Figure 2.45 shows the rainfall variation in Burkina Faso. It is estimated that between 1920 and 1995 Burkina Faso's water table dropped by 20 m. This has major implications for the growth of vegetation and the formation of soils. As the vegetation struggles to survive in the drought conditions, the soil becomes exposed and the topsoil (containing the vast majority of the fertility) is blown away by the wind or washed away by flash floods caused by tropical downpours.

The unpredictability of the Sahel climate is illustrated by the fact that Burkina Faso's cotton crop was destroyed by flash floods in 1994, and in 1996–1997, 67 000 tonnes of emergency food aid were required by Burkina Faso to deal with food shortages caused by drought.

Human factors

Improved medical care resulting in reduced death rates, coupled with high birth rates mean that rapid **population growth** is occurring in the Sahel, as shown in Figure 2.46. With more mouths to feed, greater pressure is placed on the finite resources available to produce more food. In countries such as Burkina Faso, Chad, Mali and Mauritania populations are increasing at a rate of 2.5% per year (while at the same time food production increases at only 1% per year). This has contributed to a number of factors which have caused the desertification of the Sahel:

Overcultivation

Arable farmers are forced to increase yields from their land. As a result, the fallow periods allowed for the soil to regenerate are reduced and the soil becomes depleted of its nutrients. With rainfall in the region increasingly unreliable, crops often fail and the soil is degraded. Rising populations have also forced farmers to increase the amount of land utilised for arable farming. This has led to the cultivation of marginal areas e.g. on the fringes of the Sahara. These areas are not a viable option in terms of sustained crop production and as a result cultivating them accelerates the process of desertification.

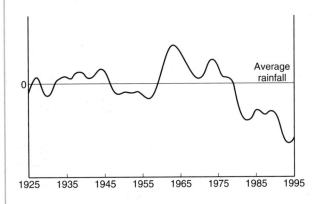

Figure 2.45   Rainfall variation in Burkina Faso.

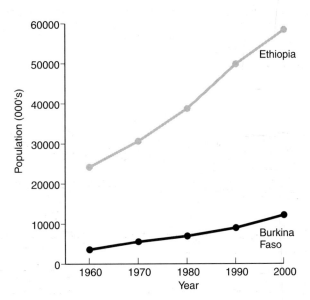

Figure 2.46   Population growth in the Sahel.

Overgrazing

The rise in population has brought about an increase in the size and number of the herds grazing in the Sahel. This places greater pressure on grazing areas particularly around the wells, boreholes, lakes and rivers where the animals are taken to drink. The concentration of herds within these areas results in the vegetation being stripped down to the roots, leaving soil exposed. In addition, compaction of the soil by trampling reduces the infiltration capacity of the soil and increases run off (and therefore water erosion). As a result farmers are forced to graze increasingly marginal areas. Land degradation follows rapidly allowing the desert to expand (Figure 2.47).

Figure 2.47   Forest land turned to desert by over-grazing and deforestation.

Deforestation

Rising populations place an ever increasing burden on woodland which constitutes the main source of domestic fuel in the Sahel. Since 1900, 90% of forests have been cleared from the Ethiopian Highlands. Deforestation removes the binding effect of tree roots, the protection afforded by the foliage and the ability of the trees to reduce wind speeds (Figure 2.47). The shortage of **fuelwood** means that more and more families are having to resort to animal dung and crop residues for their domestic energy requirements. The dung and crop residues would normally be used as fertiliser so the soil is further degraded as it is deprived of essential nutrients. In the long term, this will affect crop yields.

Urbanisation

In many parts of the Sahel there has been a rapid growth in **urban populations** as people move from rural areas in search of jobs, a higher standard of living and better education opportunities. The population of Niger's capital, Niamey, rose from 207 000 in 1980 to just under 1 million in the year 2000. The growth and development of small settlements in rural areas has also increased, as schools have been set up e.g. Korr in northern Kenya. All these settlements create a growing demand for firewood, so large areas of forest are cut down and the wood sold in the settlements. Much of this fuel is sold as charcoal because it is lighter, not as bulky and therefore cheaper than wood to transport. The wood is transformed into charcoal by burning it very slowly in earthen pits. This process is extremely inefficient and half the energy of the wood is lost during conversion. It is estimated that 1.3 billion people worldwide are consuming fuelwood resources at a rate which is not sustainable.

Growth of cash crops

During the twentieth century some farmers in the Sahel moved away from traditional techniques and began growing cash crops such as cotton and rice. This resulted in monoculture, which rendered the soil infertile. This was often accompanied by inappropriate farming techniques such as flood irrigation. In Niger, for example, water is pumped out of the River Niger to irrigate fields of rice. The flood irrigation can lead to salinisation of the soil and as a result it becomes degraded.

Questions

4 'Urbanisation is the main cause of land degradation in the Sahel.' Do you agree or disagree with this statement? Justify your answer.

5 The promotion of cash crop agriculture presents a conflict of interests for countries in the Sahel. Discuss the arguments for and against increasing the cropland area given over to cash crops.

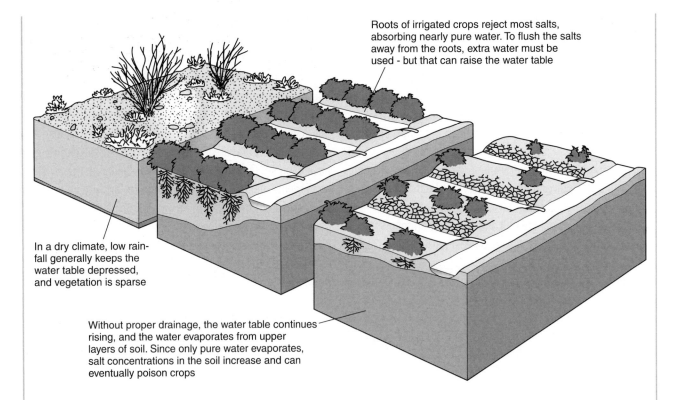

Roots of irrigated crops reject most salts, absorbing nearly pure water. To flush the salts away from the roots, extra water must be used - but that can raise the water table

In a dry climate, low rainfall generally keeps the water table depressed, and vegetation is sparse

Without proper drainage, the water table continues rising, and the water evaporates from upper layers of soil. Since only pure water evaporates, salt concentrations in the soil increase and can eventually poison crops

Figure 2.48   How irrigation can cause salinisation of the soil.

Consequences of land degradation

Physical impacts

The increasing frequency of drought periods, the farming of marginal areas along the northern boundary of the Sahel and deforestation have led to the southward expansion of the Sahara desert. In the last 50 years, 65 million hectares of the Sahel have turned to desert and in Sudan the Sahara has advanced by 100 km in 17 years. Desertification is now estimated to be spreading at a rate of 1.5 million hectares per year in the Sahel.

Rills and gullies

The exposure of soil leaves it open to the intense tropical downpours of the ITCZ. As a result overland flow removes topsoil and cuts into the land forming **rills** and **gullies**. This is also a problem where soil has been compacted by the trampling of animals.

Loss of topsoil

Strong winds, such as the Harmattan, erode the topsoil from land which has become degraded due to overcultivation, overgrazing and deforestation. During the drought of the 1970s, three times the normal amount of topsoil was blown away. Some of this was deposited in Barbados nearly 5000 km away.

Salinisation

Inappropriate farming technology and the growth in cash crop farming has led to an increase in flood irrigation, causing salts to accumulate in the soil (Figure 2.48). If irrigated soils are not drained adequately, salts will build up in the root zones of crops. This can kill the crops and contaminate the soil. To reverse the process and rejuvenate the soil is very expensive.

Social and economic impacts

Malnutrition and starvation

The failure of crops year after year leads to **starvation** and death e.g. Sudan and Ethiopia in the mid 1980s. **Malnutrition** may also be attributed to the reduction in the range of crops grown in order to concentrate on cash crops.

Migration

The degradation of farm land and the increased pressures created by rising populations have forced people to **migrate** away from their home areas in the Sahel. This has led to the loss of traditional farming techniques and in some cases, e.g. northern Niger, a demographically imbalanced rural population. This can occur when younger men move away in search of an alternative income, leaving the women, children and the elderly at home.

Dependence on external support

Where the effects of land degradation and drought have been most acute, people have come to rely on food and other aid sent from other countries, e.g. the Band Aid campaign of 1983. This can lead to **over dependency** on external help.

Access to education

In many of the Saheian countries such as Burkina Faso, education and health care must be paid for and therefore the loss of income brought about by the failure of crops and herds may mean students have their schooling interrupted and individuals go untreated when ill.

The most obvious economic impact of rural land degradation in the Sahel is the loss of income to farmers. In addition, as the yields of cash crops decrease because of the exhaustion of the soil, the quantities of commodities such as cotton being exported will fall and countries will lose a valuable source of foreign currency.

Solutions to land degradation

Afforestation

Planting new trees helps to bind the soil, and also provides shade, windbreaks (reducing wind erosion), nutrients and fuel for families (Figure 2.50). Depending on the species planted the trees may also provide nuts and fruit for humans and animals. One important element of any afforestation programme is to ensure the supply of fuelwood is sustainable. This has led to the practice of **agroforestry** (Figure 2.51) where farmers utilise the benefits offered by the trees for their crops or grazing their animals. The trees help shade crops and animals from the harsh sunlight and provide an additional source of nutrients for the crops (in the form of decaying leaf matter).

In parts of West Africa, the Kad tree, a variety of Acacia, has been used in an attempt to restore soil fertility and productivity (Figure 2.49). The advantages of the Kad are that:

- it retains its leaves in the dry season, so providing shade for the soil
- it fixes nitrogen in the soil, thus raising fertility
- its seeds and pods provide a rich source of protein for cattle and goats.

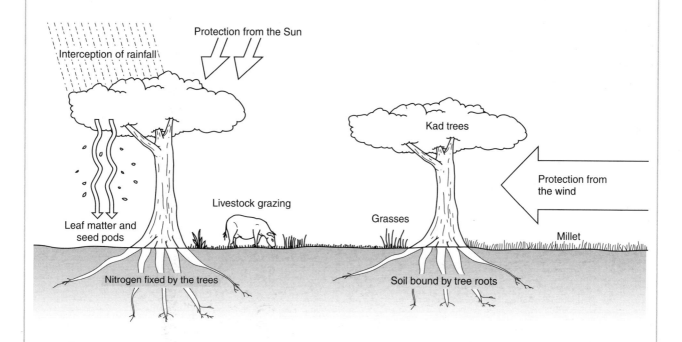

Figure 2.49   Agroforestry techniques.

Figure 2.50   Farmland in Burkina Faso, recovered using afforestation.

Figure 2.51   Agroforestry in Mali.

Stone lines

The construction of low lying **stone lines** (or Diguettes) along the contours of the land is a very effective way of reducing soil erosion and the loss of water by overland flow (Figure 2.52). In addition it can be completed as part of a community project with families helping each other at very little cost (assuming that a source of stones is available). Communities in Mali, Sudan, Burkina Faso (Figure 2.53) and Niger, with the support of aid organisations such as Oxfam and Tear Fund, have used this method very effectively. In some cases crop yields have been increased by as much as 50%.

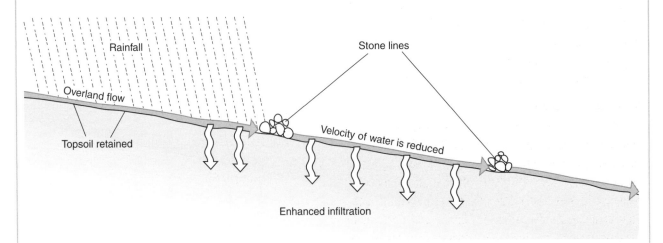

Figure 2.52   The technique of using stone lines to prevent soil erosion.

Figure 2.53   Stone lines in Burkina Faso.

Managed grazing areas

Research and tests have shown that grasses are able to re-establish themselves very quickly given an undisturbed fallow period in which to recover (1–2 years). Therefore if areas are fenced off from grazing animals on a rotational basis it is possible to sustain the herds without long term damage to the soil. The main disadvantage of this method is the cost of fencing and the management of the grazing areas e.g. around the settlement of Korr in Northern Kenya.

Reduced herd sizes

Reducing the numbers of cattle, sheep and goats and focusing on the quality rather than quantity of the herds, decreases grazing pressure on the land. It also reduces the impact of soil compaction which causes increased overland flow and reduced infiltration. This is a very culturally sensitive issue as traditionally the size of a herd is often a measure of an individual's status within his tribe and therefore herders may be resistant to change.

Education

Teaching farmers about the causes and consequences of rural land degradation can help change attitudes. An example of this is the importance of the appropriate irrigation techniques which avoid the salinisation of the soil e.g. drip irrigation. Unfortunately these alternatives are often more expensive.

Fuel efficient stoves

The development of stoves like the **Jiko**, which is cheap and uses approximately half the quantity of fuelwood required by an open fire can play a part in reducing rural land degradation. If these stoves are used, the demand for fuelwood is reduced and trees will be retained for longer. This technique can be particularly effective in areas where the cost of alternative fuels, such as Kerosene, are beyond the means of local people.

Case Study — Livestock project in Burkina Faso

Background

Burkina Faso has a population of 9.4 million, and this is increasing at a rate of 2.5% per annum. The location of Burkina Faso is shown in Figure 2.54. Life expectancy is 48 years and approximately 85% of the population live in rural areas. Adult literacy is low (8.4% for females and 27.9% for males).

Burkina Faso faces a number of fundamental problems which include:

- desertification and land degradation caused by over use of natural resources and a series of drought years
- the challenge of feeding a rapidly increasing population while relying heavily on basic farming methods which have not changed in centuries
- government policies and growing consumerism which have led to a growing pressure to grow cash crops
- a heavy national debt problem
- AIDS is a major threat and is already affecting the balance of the population and reducing debt repayment.

This livestock project is based in the Tangaye, 140 km north of Ouagadougou (Figure 2.55), an area farmed by the Mossi tribe. Tangaye is an agro-pastoral zone where subsistence farming is the main livelihood. Land degradation and inadequate rainfall have made it increasingly difficult for families to produce sufficient food. One of the main consequences of these factors has been a significant rise in rural–urban migration and the subsequent breaking up of rural communities. The project is supported by the British charity Tear Fund.

Aims of the project

The initial aims of the project were to provide improved living standards and greater food security for the families in Tangaye. This was to be achieved through provision and training in improved agricultural techniques.

The objectives were to:

- increase food production and diversify crops
- improve soil fertility through the use of animal manure

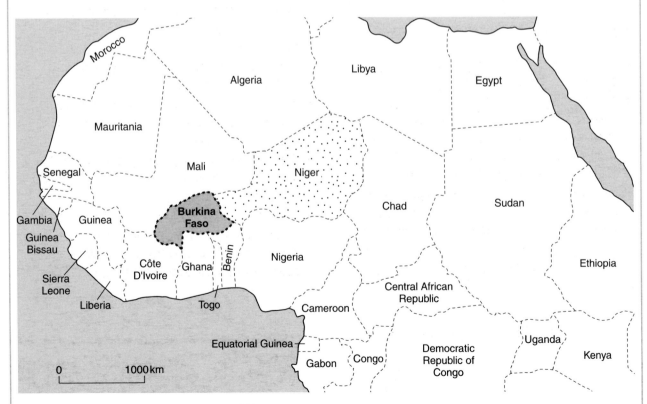

Figure 2.54 Location map of Burkina Faso and Niger.

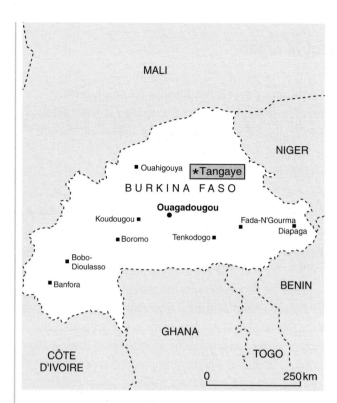

Figure 2.55   Location of Tear Fund project in Burkina Faso.

- reduce the burden of work on families, especially women, and increase their productivity through the use of appropriate technology.

Methods used

- Education and training was provided for farmers in the selection and care of their oxen.
- The establishment of manure-producing cowsheds. These enable the animals to be well looked after and at the same time produce a reliable source of manure which can be used to improve soil fertility. The cowsheds were to be paid for by the farmers, although other equipment such as feeding troughs, spades and scythes were funded by Tear Fund. The manure is produced in a pit (4 m diameter and 4 m deep). During the dry season, water is added to the pit to ensure the manure rots fully. This is important because if the

manure has not rotted sufficiently, it can actually stunt plant growth.

- Purchase of cattle (one per family).
- Purchase of agricultural equipment, in particular ploughs. By using cattle to do the heavy work on the land, farmers are able to cultivate a larger area more efficiently. In order to cultivate an area of 100 hectares, 20 cattle and 20 ploughs are needed.
- Provision of veterinary care. The cattle involved in the project were vaccinated in order to ensure they were healthy enough to carry out the work necessary.
- Stone lines were constructed along contours to retain water, decrease run off and increase infiltration.
- A community approach was engendered. The farmers were arranged into groups to encourage mutual support. In addition a system was set up whereby after three years families included in the project agree to sell their cattle and buy three younger animals. Of the three purchased, two are retained and the third is given to another family.
- Some reforestation was also carried out.

Outcomes

- Families are able to cultivate between two and three times the area of land they could by hand.
- Increased food production. It is estimated that crop yields are around 25% higher when manure is used.
- Other families in the area are now copying the techniques introduced. The number of cattle has doubled in the last five years.
- Improved standards of health, particularly amongst women as they are able to use the animals for transport and are saving time and energy.
- The main problem facing the project is enabling it to become self-sufficient and therefore no longer reliant on outside agencies for funding.
- There have also been a small number of deaths amongst the cattle which obviously inhibits the effectiveness of the project and highlights the need for adequate veterinary care to be available and accessible to the locals.

Case Study

Integrated development project in Niger

Background

This project was established to help the Tuareg ethnic group in Niger (Figure 2.54). The Tuareg are semi-nomadic pastoralists whose methods have changed little over the centuries. In recent years they have experienced increasing hardship through the loss of large numbers of livestock. This has been as a result of drought, rising population, desertification and the encroachment of settled agriculture.

The focus of the project was in an area north east of Abalak (Figure 2.56). It is an area of semi-arid grasslands, with a total annual rainfall of around 250 mm. Three fossilised valleys dissect the region, each with an intermittent river during the rainy season. The area has been affected by severe droughts in 1993–1994 and 1997–1998. Most of the soils are loose, sandy and retain very little moisture. However, the soils on the floodplains of the valleys have a higher clay content and are agriculturally viable. Large areas of natural forest have died out since 1984. This has left the soil exposed, erosion has followed and any natural regeneration of the forest is impeded by animals.

Herbaceous plant species found in the area provide excellent fodder but locals have reported at least twelve species that have disappeared since 1984. The principal products of the Tuaregs' animal husbandry are milk, cheese, leather, skins and meat.

Aims

The main objective of the project was the establishment of **fixation points** i.e. fixed areas where the Tuareg can spend most of the year. The aim of these fixation points is to 'provide the population with the ability to improve, manage and use the available natural resources, and survive droughts and famines.' Secondary to this is the hope that fixation points will provide the opportunity to develop and integrate improved education and health services, while still retaining local culture and tradition. It is important to emphasise that this is not **sedentarisation** i.e. the permanent settlement of people in one place. The Tuareg still migrate, but the fixation point develops a central area where they already spend the majority of the year.

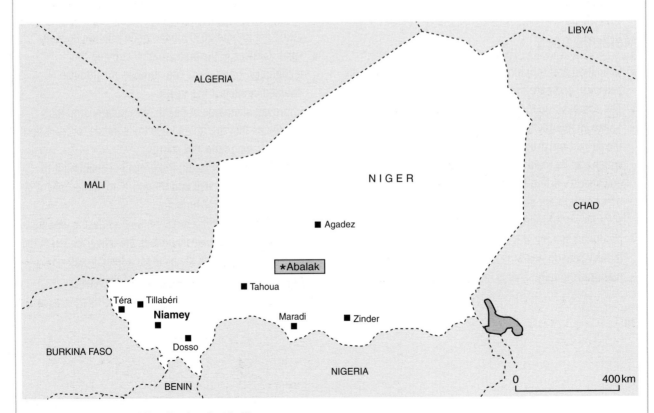

Figure 2.56   Location of Tear Fund project in Niger.

Methods

- Stone lines were constructed to improve water retention and infiltration.
- The creation and repair of wells, five wells have been created and five repaired.
- The cutting and selling of deadwood, this provides additional income but must be sustainable.
- Traditional migration during the rainy season is encouraged.
- Training in the production of mud bricks and woodless construction techniques. Communal house building events are organised. The aim is to reduce their reliance on wood for this purpose.
- Introduction of better quality breeds of livestock. Herds are traditionally large, inefficient and unproductive. The introduction of alternatives means herd sizes can be reduced.
- Economic diversification – the people are being educated about alternative sources of income as herding alone cannot support the increasing population. This is a long term goal of the project.
- Pasture management – this includes the fencing off of high quality pasture, the rotation of livestock and regeneration of pasture, so that everyone benefits.
- Afforestation – tree nurseries have been established and from this a tree planting programme has been developed.
- Provision of tools and seeds – the aim here is to encourage the Tuareg to cultivate crops in addition to rearing their herds. This in turn provides them with an alternative source of food and reduces their reliance and vulnerability to market forces.

Outcomes

- It is estimated the project has benefited around 2500 people directly and about the same number indirectly.
- The stone lines have led to the regeneration of some herbaceous plants, the establishment of pasture and the cultivation of crops such as sorghum, millet and cowpeas. The construction of one stone line can regenerate at least 20 hectares of land.
- The Tuaregs' food supply is now more diverse and reliable.
- Afforestation is now taking place as a result of the seedling nurseries.
- A much stronger community approach to the problems facing the Tuareg is developing. This is a particularly significant development given the fiercely independent nature of the Tuareg.
- The traditional attitudes towards herd sizes are being changed as the people see the benefits to be gained from having a smaller, better quality herd.
- The additional income being generated by overall diversification is enabling the Tuareg to send their children to school.
- The rate of change is very slow due to the depth of the Tuaregs' traditions and culture.
- Conflicts have been experienced between the Tuareg and other ethnic groups e.g. the Fulani, who are moving into the area because of the widespread problems of rural land degradation in Niger.
- The project is still reliant on funding from outside agencies, such as Tear Fund.

Questions

With reference to the last two case studies:

8 Outline why each project was necessary – it maybe helpful to do this using a spider diagram.

9 Are the techniques used in each project appropriate to the area they serve? Justify your answer.

10 Draw up a table of the main benefits and drawbacks of each project.

11 Using the relevant information from the last two case studies, draw up two revision tables using the headings given below, and complete them – one for the Sahel and one for the Great Plains. Remember to record as many facts, figures and examples as you can.

Causes of rural land degradation	Consequences of rural land degradation	Solutions and their effectiveness

12 'Climatic factors are the main cause of rural land degradation.' Do you agree with this statement? Support your answer using examples from either North Africa or North America.

13 Describe in detail the impact that land degradation has had on the **physical landscape** of the Sahel.

14 Explain, giving examples, why many of the techniques used to combat land degradation in North America would be considered inappropriate in the case of the Sahel.

15 'Water erosion has a more devastating impact in terms of land degradation than wind erosion.' Discuss the validity of this statement, using examples from the Sahel and North America to support your answer.

Key terms and concepts

Summary

Having worked through this chapter, you should now know:

- the nature and processes of soil erosion and land degradation
- the effects of climate and its variability on these processes within North America and North Africa
- that ecosystems are modified by human activity
- the physical, social and economic impact of soil erosion and land degradation
- specific examples of soil conservation and land management strategies and their effectiveness through case studies from North America and Africa.

Urban Change and its Management

Links to the core

The movement of population from rural to urban areas has now become the world's biggest migration pattern. As a consequence an increasingly large proportion of the world's population is now classified as urban. This section will build on the ideas and concepts introduced as part of the urban geography section of the core unit: urban systems, urban functions and structures and urban change. It examines the phenomenon of urbanisation and focuses particularly on the problems of managing the changes which face growing urban areas. Relevant aspects of other parts of the human core such as population geography and industrial geography have been fully integrated into the case studies. Two cities will be studied in depth: Toronto in Canada, exemplifying managing change in a rapidly growing city in a More Economically Developed Country (MEDC), and Karachi in Pakistan, an expanding city in a Less Economically Developed Country (LEDC).

Population growth and urbanisation in a MEDC

Population growth in Canada

Canada is the world's second largest country in area (9 970 610 km²) and is divided into a number of provinces and territories (Figure 3.1). The 1996 Census counted 28 846 761 people in Canada, just over half the population of the UK. The Canadian population doubled in the 45 years after 1951 and although the rate of growth slowed between 1991 and 1996 (Table 3.1), Canada's population continues to grow at the fastest rate of all the G8 industrialised nations (Table 3.2).

The distribution of Canada's population is also significant. Over 80% of the Canadian population is concentrated into a narrow strip a few hundred kilometres wide along the length of the USA–Canada border. As Figure 3.2 shows, despite its vast area, this concentration of population indicates a high degree of urbanisation.

Urbanisation in Canada

Most of the rapid expansion of Canadian cities has taken place since 1950. This contrasts sharply with the experience of many UK cities which have experienced major losses of population due to the process of counter-urbanisation during the same period.

Figure 3.1 Canadian provinces and territories.

	Total Population	Population Increase	Growth Rate %
1951	14 009 429		
1956	16 080 791	2 071 362	14.8
1961	18 238 247	2 157 456	13.4
1966	20 014 880	1 776 633	9.7
1971	21 568 311	1 553 431	7.8
1976	22 992 604	1 424 293	6.6
1981	24 343 181	1 350 577	5.9
1986	25 309 331	966 150	4.0
1991	27 296 859	1 987 528	7.9
1996	28 846 761	1 549 902	5.7

SOURCE: WORLD POPULATION (1996),
UNITED NATIONS, POPULATION DIVISION

Table 3.1 Population growth in Canada since 1951

	Average Annual Growth Rate (%)
World	1.5
G8 Countries	
United States	1.0
Germany	0.6
France	0.5
United Kingdom	0.2
Japan	0.2
Italy	0.1
Canada (1991–1996 Census)	1.1
Russian Federation	0.7

SOURCE: WORLD POPULATION (1996),
UNITED NATIONS, POPULATION DIVISION

Table 3.2 Population growth rates in G8 nations 1990 to 1995

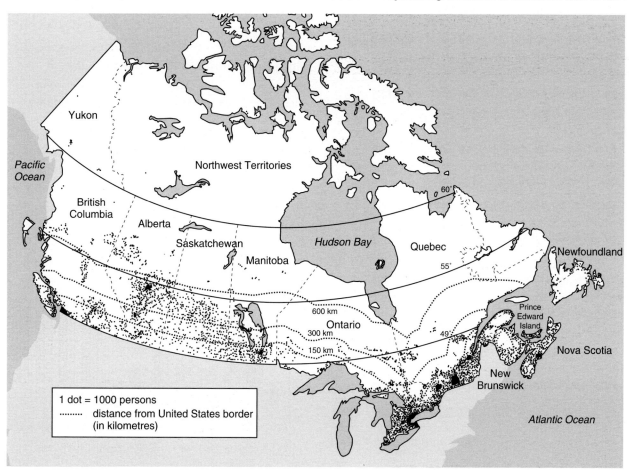

Figure 3.2 The distribution of population in Canada.

Rank in 1996	City	Population		
		1991	1996	% change
1	Toronto (Ont.)	3 898 933	4 263 757	9.4
2	Montréal (Que.)	3 208 970	3 326 510	3.7
3	Vancouver (B.C.)	1 602 590	1 831 665	14.3
4	Ottawa (Que.)	941 814	1 010 498	7.3
5	Edmonton (Alta.)	841 132	862 597	2.6
6	Calgary (Alta.)	754 033	821 628	9.0
7	Québec (Que.)	645 550	671 889	4.1
8	Winnipeg (Man.)	660 450	667 209	1.0
9	Hamilton (Ont.)	599 760	624 360	4.1
10	London (Ont.)	381 522	398 616	4.5
11	Kitchener (Ont.)	356 421	382 940	7.4
12	St Catharines – Niagara (Ont.)	364 552	372 406	2.2
13	Halifax (N.S.)	320 501	332 518	3.7
14	Victoria (B.C.)	287 897	304 287	5.7
15	Windsor (Ont.)	262 075	278 685	6.3
16	Oshawa (Ont.)	240 104	268 773	11.9
17	Saskatoon (Sask.)	210 949	219 056	3.8
18	Regina (Sask.)	191 692	193 652	1.0
19	St. John's (Nfld.)	171 848	174 051	1.3
20	Sudbury (Ont.)	157 613	160 488	1.8
21	Chicoutimi – Jonquière (Que.)	160 928	160 454	−0.3
22	Sherbrooke (Que.)	140 718	147 384	4.7
23	Trois-Rivieres (Que.)	136 303	139 956	2.7
24	Saint John (N.B.)	125 838	125 705	−0.1
25	Thunder Bay (Ont.)	124 925	125 562	0.5

SOURCE: STATISTICS CANADA

Table 3.3 The largest 25 urban areas in Canada.

For example, the population of Greater London fell by 200 000 between 1951 and 1961, and there was a further loss of 540 000 between 1961 and 1971. Liverpool, Manchester and Newcastle each lost around 18% of their respective populations during the 1960s. The 1981–1991 period in the UK revealed that although population decentralisation continued, there were some interesting exceptions. The major metropolitan centres of Manchester, Liverpool, Sheffield, Newcastle, Birmingham, Leeds and Glasgow showed continually high rates of out-migration, but Inner London experienced considerable population growth, as have several non-metropolitan urban areas.

The three largest Canadian cities have experienced very rapid growth rates post-1950 and have at least doubled their populations. Toronto quadrupled its population during this time and in 1976 overtook Montreal as Canada's largest city. During the same period, Vancouver established itself as the third largest city in Canada. The 1996 Census revealed that the population increase in the 25 **Census Metropolitan Areas** (CMAs) was 1.3 million, representing 80% of the total Canadian population growth (Table 3.3). A CMA consists of a very large urban core together with adjacent urban and rural areas that are economically and socially related to the core. The urban core has a minimum population of 100 000.

The fastest growing cities were Vancouver (+14.3%), Oshawa (+11.9%), Toronto (+9.4%) and Calgary (+9%). More than a third of Canada's population lived in the four most populous metropolitan areas:

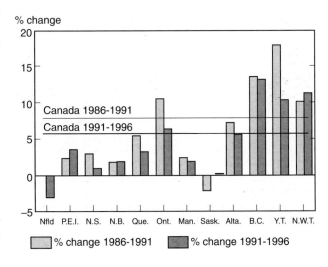

Figure 3.3 Population growth rate in Canada, 1986–1996.

Toronto, Montreal, Vancouver and Ottawa. The 1996 Census also confirmed a trend seen for a number of years showing the distribution of Canada's population gradually shifting from east to west. In Ontario, the population increased by 6.6% of which 60% was accounted for by international migration and the remainder by natural increase and migrations from other Canadian provinces. About 80% of Ontario's growth occurred in its CMAs and half of the entire population growth was within the Toronto CMA. The Toronto CMA became the first in Canada to be populated with over 4 million people.

There are three main causes of rapid urbanisation in Canada. First, the cities have attracted substantial numbers of migrants from rural areas. Figure 3.3 and Table 3.4 show a clear trend where the bigger urban areas have experienced growth, whilst those more dependent on primary industries such as Nova Scotia and Saskatchewan show the lowest population growth rates. Newfoundland, hit by the collapse of its fisheries in the early 1990s showed its first recorded loss of population as migrants left for other parts of the country.

Second, the largest Canadian cities have traditionally been the destination of overseas immigrants settling in Canada. Although the source of immigrants to Canada has changed; during the 1950s, 60s and 70s immigrants came mainly from Europe, whereas today they are mainly from Asia, the trend for them to settle, initially at least, in the main urban centres has continued. International migration accounted for over half of the population growth recorded in Toronto between 1991 and 1996. These immigrants, often young married couples, are responsible for the third cause of population growth by generating higher birth rates for the cities than for other parts of Canada.

Urban trends in Canada

Although each Canadian city has responded to the demands of population growth in different ways, it is possible to identify a number of common trends:

- the outward physical spread of each city in response to demands for low density suburban housing on the outskirts

	1996	1991	% Change	Absolute Change
Canada	**28 846 761**	**27 296 859**	**5.7**	**1 549 902**
Newfoundland	551 792	568 474	−2.9	−16 682
Prince Edward Island	134 557	129 765	3.7	4 792
Nova Scotia	909 282	899 942	1.0	9 340
New Brunswick	738 133	723 900	2.0	14 233
Quebec	7 138 795	6 895 963	3.5	242 832
Ontario	10 753 573	10 084 885	6.6	668 688
Manitoba	1 113 898	1 091 942	2.0	21 956
Saskatchewan	990 237	988 928	0.1	1 309
Alberta	2 696 826	2 545 553	5.9	151 273
British Columbia	3 724 500	3 282 061	13.5	442 439
Yukon Territory	30 766	27 797	10.7	2 969
Northwest Territories	64 402	57 649	11.7	6 753

SOURCE: STATISTICS CANADA

Table 3.4   Population growth from 1991 to 1996 in Canadian provinces.

Figure 3.4   Canadianville.

- pressures on agricultural land immediately outside the city limits for change of use to housing, light industry and office parks
- heavy pressures on the network of highways and public transport systems as a result of the movement of population, particularly at peak hours
- major changes in land use within the cities in response to changing requirements of commerce and industry.

Also, in terms of physical appearance, there is an increasing similarity between Canadian cities. This is best represented by a study of their profiles which have assumed a pyramid or inverted cone shape (Figure 3.4). The apex is located at the heart of the **Central Business District** (CBD) which is usually the site of the tallest building or skyscraper. With increasing distance away from this point, building height decreases. This classic profile can be explained in terms of land values and rents. Economic and competitive market forces have created the central skylines of Canadian cities. Despite changes in central areas of European cities in recent years, it is true to say that their skylines are often dominated by churches, cathedrals and castles and it is certainly not possible with any certainty to point from a distance and identify their peak land values. In Canadian cities this can be done with monotonous ease.

Questions

1 Explain why 'Canada is more urbanised . . . than many Western European countries.'

2 Outline the differences in the pattern of urban growth in Canada between 1951 and 1981 compared with that which took place in most Western European countries during the same period.

3 On an outline map of Canada locate and mark the 25 largest cities.

4 What were the reasons for the rapid growth of Canadian cities between 1951 and 1991?

5 a Explain why population growth from 1991 to 1996 was greatest in Ontario, Alberta and British Columbia.

   b Account for the high rates of growth in Yukon and the Northwest Territory during the same period.

6 Describe and explain the significance of the Canadianville Model.

7 Explain why competition for land is greatest in the Central Business Districts of Canadian cities?

Case Study   Toronto

Situation, site and growth

The Huron Indians called the present site of Toronto 'the carrying place' because it was the centre of the area's fur trade. The area was strategically important in controlling major trade routes and first the French, and later the British built a series of forts in the area (Figure 3.5). In 1788 the British purchased control of the area from the Mississauga Indians paying 146 barrels of blankets, axes and cloth. The first lieutenant governor of Upper Canada, John Graves Simcoe, had his doubts about the value of the purchase and commented that 'the city's site was better calculated for a frog pond or a beaver meadow than for the residence of human beings'. Today, the land is worth many billions of dollars and is home to over 4 million people.

Originally named York, the population reached 1000 by 1818, and in 1834 the town became a city and was renamed Toronto. The city grew quickly during the nineteenth century mainly due to the advantages of its site and situation. The **site** i.e. the actual piece of ground on which the city has developed, was chosen because of its large, natural harbour on Lake Ontario sheltered by a peninsula which later became the Toronto Islands (Figure 3.6). The **spit**, formed from clay, sand and gravel, was separated from the shoreline in 1858 following a major storm and the hooked shape of the islands creates a natural harbour.

Many features of the site of Toronto show the influence of glaciation. At the end of the ice age Lake Ontario was much bigger than today and was called Lake Iroquois. Meltwater was trapped behind ice which blocked the natural exit channel down the St Lawrence Valley, so instead water flowed southwards and created the Hudson-Mohawk routeway to New York.

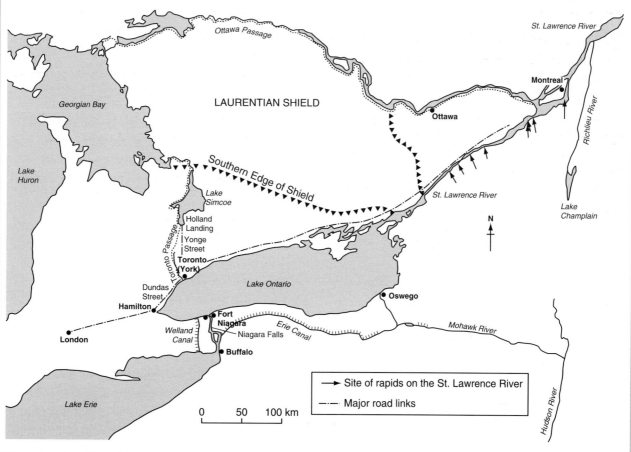

Figure 3.5   Location of Toronto.

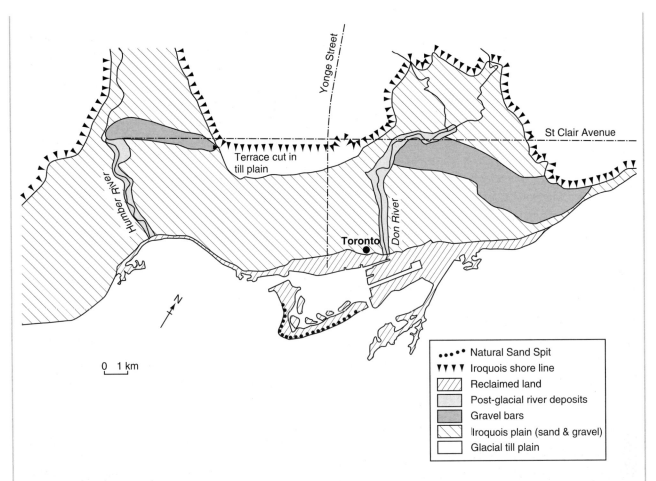

Figure 3.6 Physical site of Toronto.

The old Lake Iroquois shoreline in Toronto is marked by a steep **bluff** or slope between 20–25 m in height (Figure 3.7). The construction problems caused by the bluff meant that much of Toronto's early growth took place east-west rather than towards the north. South of the bluff the Iroquois plain allowed easy construction along with a ready supply of building materials from the sands and gravels deposited along it. Extensive gravel bars were deposited where the Humber and Don Rivers entered the former Lake Iroquois (Figure 3.6).

The site of the city is greatly influenced by the valleys of a number of rivers and creeks which flow into Lake Ontario (Figure 3.7).The most significant of these are River Humber, River Don, Highland Creek and the Rouge River. These rivers and their tributaries have cut deep **ravines** into the **till plain** deposited by glaciers, and have restricted urban development to the land in between them. The ravines themselves are unsuitable for building because of their low-lying nature and also because much of their surface material is unconsolidated.

In October 1954 Hurricane Hazel hit the city and 36 people were killed where housing had been built on the Humber floodplain. Bridges have had to be built to allow east-west routeways across the modern city, although the valleys themselves have been used for transportation corridors (Figure 3.8) and some form extensive parklands (Figure 3.24).

The original old town was laid out on the lake plain. As transportation networks developed during the second part of the nineteenth century, Toronto began to expand rapidly. Industrial growth was stimulated by the railways and the development of the port.The construction of the Welland and Erie Canals improved the city's links with New York via the Mohawk and Hudson Rivers. The coming of the railways reinforced Toronto's position as the hub of transportation with improved links with the rest of Canada and the north eastern USA. This early growth took place to the south of the core area. As industry developed and more European immigrants arrived in search of jobs, housing was largely constructed to the east and west of the original site of the city.

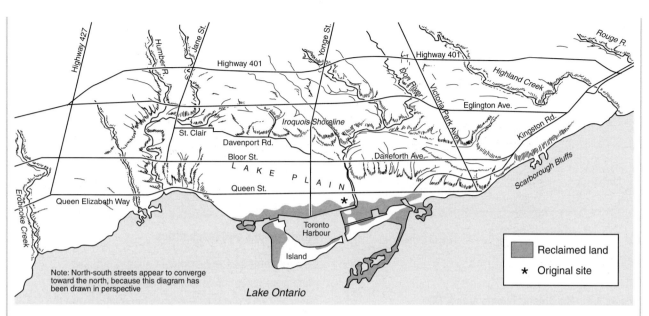

Figure 3.7   The topography of the City of Toronto.

These houses were mainly three-storey semi-detached on 5 m wide plots. They were built at such a high density that by 1945 the city had one of the highest population densities in North America.

By the early 1950s, with continuing migration from other parts of Canada and increased immigration, the city was expanding rapidly and the urban structure developing to include neighbouring municipalities.

Figure 3.8   The Don Valley expressway.

Questions

8 Outline both the advantages and disadvantages of the site of Toronto for urban growth.

9 Using an atlas and Figure 3.5 explain why Montreal was better located for the development of international trade than Toronto, before the mid-nineteenth century.

10 Explain the advantages to Toronto of the construction of the Welland and Erie Canals.

11 In what ways was continued European immigration vital to the development of Toronto in the late nineteenth and early twentieth centuries?

Population distribution and growth

As well as being Canada's largest city, Toronto is now the eighth largest urban centre in North America. In 1996 the Toronto Census Metropolitan Area was home to 4 263 757 people, a growth of almost 10% since 1991. The Greater Toronto Area (GTA) (Figure 3.9) includes the City of Toronto itself and the regions of Halton, Peel, York and Durham. The population of the city area was 2.385 million in 1996 accounting for 51.5% of the GTA population.

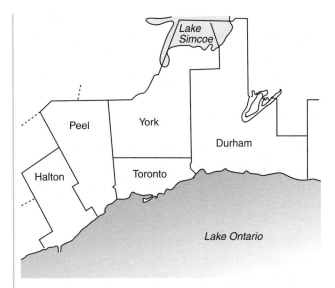

Figure 3.9   The Greater Toronto Area (GTA).

Most of the recent growth has taken place in the outlying regions where the rate of population increase is three times that of the city itself. Toronto is much more compact than the GTA regions, in 1996 Toronto's average population density was 3786 people per km², compared to a combined average of 1571 for the regions. Densities in Toronto's urban core reach as high as 6740 people per km². The main reason for the lower density in the outer area is the desire of many people to purchase single-family homes on large lots and land is much cheaper further away from Toronto. There is in fact almost no open land remaining within the city boundaries.

Indications are that growth will continue to be greater in the regions and by 2011 it is estimated that 55% of the GTA's population will live outside the city. The populations of Peel and York are expected to double in the next 20 years. Much of this growth is due to the availability of jobs with almost 50 000 in government alone in Toronto.

Toronto's population, like that of the country as a whole, is ageing. By and large Toronto is ageing and older, whilst the surrounding regions are younger. Toronto has 52% of the working age and 63% of the senior GTA populations. Many senior citizens remain in the houses which they purchased after the Second World War and into the 1970s. Others prefer to live in the city because of transport accessibility and equally accessible social and human services. The '**greying**' of Toronto's population is reinforced by the tendency for many young families to move to larger houses in the regions around the city leaving the older population living in the city itself. Evidence of this trend is

found in the distribution of children. The regions have 52% of pre-school and 55% of school age children.

Despite the outward movement of population towards the suburbs, the population of the City of Toronto itself is expected to grow slightly due to two identifiable trends:

● new housing areas are being constructed on old **brownfield** industrial sites such as the railway lands near the lake
● young adults are moving into older areas of the city where they renovate older houses (**gentrification**).

Ethnic groups and inequalities

Toronto continues to be one of Canada's primary immigrant reception areas, and as a result the city is ethnically very diverse. The city's big magnets are the availability of rental housing and the presence of its existing immigrant communities. Between 1991 and 1996 315 460 people emigrated to Toronto. Providing low-cost housing for such large numbers is becoming increasingly difficult. Federal and provincial housing projects have been drastically cut and the number of new rental units in the city cannot meet existing or predicted demand. As a result, homelessness in Toronto has reached unprecedented levels, evidenced by the increasing numbers of people living on the streets, using public buildings as shelters and relying on food banks and other charitable services. It is estimated that in 1996 on any given night, 3200 people were using Toronto's hostels.

As indicated by Figure 3.10 the longest-established immigrants, such as those from Italy and the UK, tend to have the highest proportions already moved from the central to outlying parts of the city. In the 1950s, 75% of the City of Toronto's population was of British origin and even by 1981 the figure was 48%, by 1986 it had fallen to 28%.

Initially the UK was the main source of Toronto's immigrants, followed by Italy. More recently this has switched to the Caribbean, Vietnam, Hong Kong and India. There is a tendency for certain areas of the city to house concentrations of immigrants from one country (Figure 3.10), but overall Toronto is a very cosmopolitan city, with immigrants gradually establishing themselves and moving out from the city centre to better housing in the surrounding suburbs. Perhaps one of the best examples of this is the Chinese community which has traditionally been centred in Toronto's Chinatown located on the edge of the CBD.

'Toronto boasts about its downtown Chinatown. It is one of the multi-cultural neighbourhoods the city celebrates with special features, such as street signs in the residents' mother tongue. The area has the busy feel of a multi-ethnic street carnival. Signs and storefronts blaze with a patchwork of mandarin red, emerald green and imperial gold. Scores of restaurants venture far beyond the fried rice, chicken balls and chow mein served across North America, offering foods from every region of China. Dozens of shops display herbal remedies such as ginseng and lotus root, as well as clothing, hand-hammered woks and other housewares, furniture, ornate jewellery and Chinese books and records. The shelves and bins of grocery stores are packed with specialities such as bok choy and Chinese cabbage – now widely grown on southern Ontario farms – along with sweet, yellow Chinese pears, hot bean sauce, sticky brown rice paste and chicken feet.

Chinese people can find almost any service, or conduct almost any transaction, in their own languages. Large numbers of Chinese lawyers, doctors, herbalists and acupuncturists have set up practices. Streets, banks, police stations and most other services are identified by both English words and Chinese characters. Four theatres screen Chinese movies and there are branches of several Hong Kong banks. At night the streetscape of shop signs and store windows create a colourful tapestry of lights. The activity lasts hours past midnight: many of the restaurants stay open until four or five in the morning.'

[Farewell to Chinatown: Peter Gorrie, Canadian Geographic]

In recent years successful Chinese–Canadians have been moving uptown to Scarborough, some 25 km north of the CBD. Located amongst one of Toronto's sprawling post-war suburbs, this Chinatown is far bigger than the one in the city centre and it is mainly the young and affluent Chinese who are making the move, leaving behind an ageing population in the traditional Chinatown. The City of Toronto's Chinese population increased from 5000 in 1951 to over 350 000 in 2000 with many new immigrants now going straight to the suburbs if they can afford to. Most of those moving into the downtown quarter are from mainland China and other parts of south east Asia. They have little education and few skills and many are illegal immigrants; some are criminals who contribute to the area's high crime rate and there are now fears of a racist backlash.

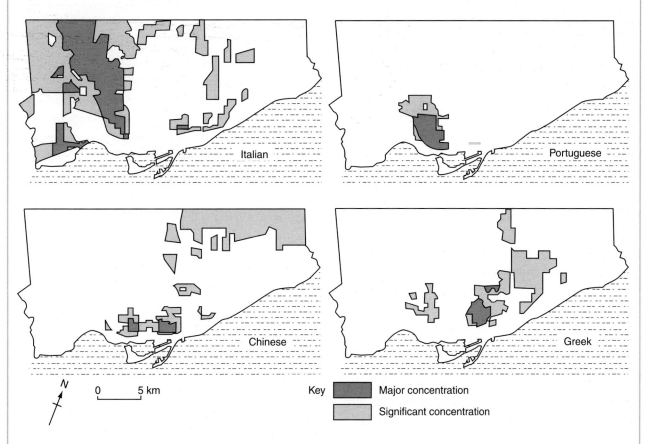

Figure 3.10   Major immigrant concentrations in Toronto.

The Chinese community have been discriminated against in Canada's past. In 1885 the federal government imposed a $50 head tax on each Chinese immigrant, which was increased to $500 by 1893. Between 1923 and 1947 legislation barred Chinese immigration into Canada and even today the Chinese are under-represented in most elected positions. The recent transition to the suburbs may herald the start of a new era for Toronto's growing Chinese community.

Questions

12 Describe and explain the growth and change in the distribution of Toronto's population since 1950.

13 Outline the problems for city planners which have resulted from the large increase in population in the suburban areas of the city, and in the surrounding regions.

14 Which groups of people are moving into the inner city area and why?

15 Using Figure 3.10 outline the evidence which suggests that the Italians were amongst the first 'non-UK' immigrants in Toronto.

16 From those shown on the maps in Figure 3.10, which nationality are the latest group of immigrants? Support your answer with map evidence.

17 Describe and explain the distribution of the Chinese population in Toronto.

18 What visual evidence might you observe on a visit to the city to support the claim that Toronto is a very cosmopolitan city?

Managing urban change in Toronto

Following the Second World War, Toronto grew very rapidly and a new form of local government structure was required to manage and plan for the changes which were taking place in its urban structure. Although the population of the City of Toronto itself was not increasing, the populations of the surrounding suburbs, each of which had its own municipal government, were exploding. With the high density of inner city housing, and the rapid growth of car ownership in the post-war years, demand for housing and schools in the suburbs was increasing at alarming rates. It was necessary to provide an overview of the entire

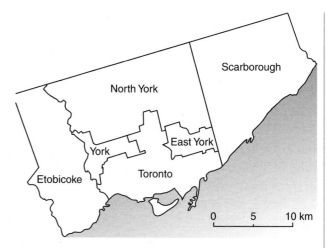

Figure 3.11   The City of Toronto.

region for planning purposes and a new system of local government was established which has since been modified several times in response to the continually changing geography of the Toronto area:

● In 1953 the City of Toronto and its 12 suburbs were amalgamated under a regional government called Metropolitan Toronto. Responsibility for strategic planning was taken over by the new Metro authority but local services such as education and many community services were retained by local councils.

● In 1967 the 13 municipalities were reduced to 6: the cities of Toronto, Etobicoke, North York, Scarborough, York and the Borough of East York (Figure 3.11).

● In 1998 these were amalgamated into one authority called the City of Toronto. The electoral wards are shown in Figure 3.12. The creation of the newly defined city in 1998 was recognition that Toronto is the centre of a broader Greater Toronto Area within which the regions of Halton, Peel, Durham and York and the City of Toronto itself are mutually dependent.

Managing land use change

As the city has grown and developed, its neighbourhoods have tended to take on specific characteristics, and to be used for specific purposes. Through an analysis of the land use pattern which has emerged (Figure 3.13) the changes and current problems facing the city can be identified.

Central Business District

The location of Toronto's CBD is shown in Figure 3.13. Toronto is Canada's leading financial centre and the greatest part of that activity takes place within the CBD.

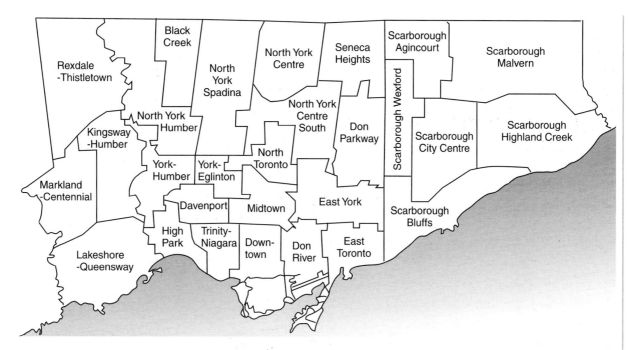

Figure 3.12 The new City of Toronto.

Toronto's CBD extends south to the railway lands, north to Bloor Street, west to University Avenue and east to Jarvis Street (Figure 3.14).

As in any large city, the CBD contains the greatest concentration of tall buildings (Figure 3.15). The nearer a building is to the CBD, the taller it is likely to be because

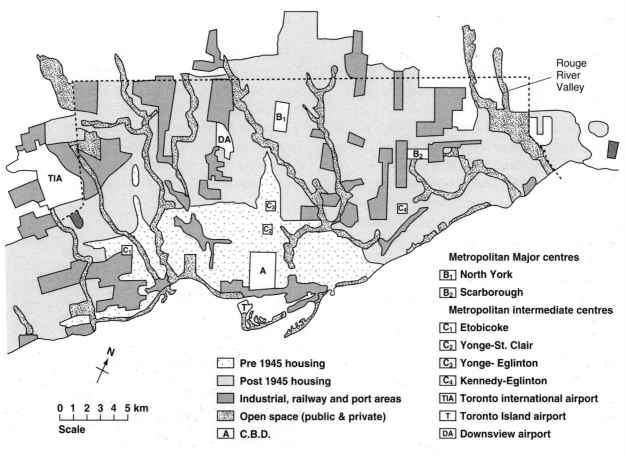

Metropolitan Major centres

B₁ North York
B₂ Scarborough

Metropolitan intermediate centres

C₁ Etobicoke
C₂ Yonge-St. Clair
C₃ Yonge- Eglinton
C₄ Kennedy-Eglinton
TIA Toronto international airport
T Toronto Island airport
DA Downsview airport

Pre 1945 housing
Post 1945 housing
Industrial, railway and port areas
Open space (public & private)
A C.B.D.

0 1 2 3 4 5 km
Scale

Figure 3.13 Land use in Toronto. C

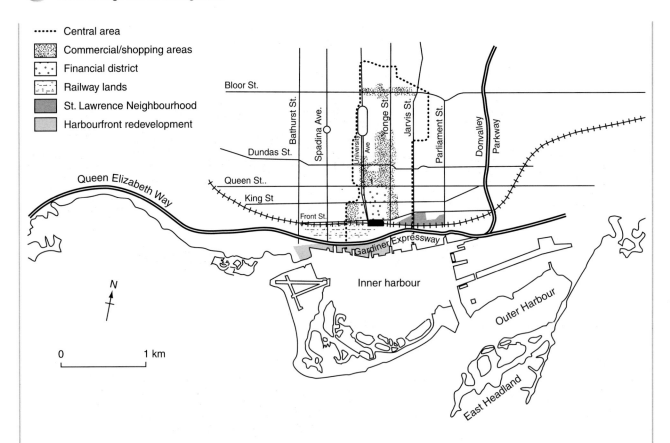

Central area
Commercial/shopping areas
Financial district
Railway lands
St. Lawrence Neighbourhood
Harbourfront redevelopment

Bloor St.
Bathurst St.
Spadina Ave.
University Ave.
Yonge St.
Jarvis St.
Parliament St.
Donvalley Parkway
Dundas St.
Queen Elizabeth Way
Queen St..
King St
Front St.
Gardiner Expressway
Inner harbour
Outer Harbour
East Headland

N

0   1 km

Figure 3.14   Toronto City centre and Harbour area.

that is where land is most expensive. The central area functions as a primary location for government, corporate head offices, financial institutions, retailing and tourist attractions. Toronto has more head offices than any other Canadian city. The highrise office towers average over 30 stories. One of the tallest office buildings is *First Canadian Place* which is 72 stories high and rises 290 m with 58 lifts. It contains thousands of legal, brokerage, banking and other business offices. On an average working day the complex is occupied by approximately 30 000 people whose total annual income exceeds a billion dollars. These people need to work in close proximity for meetings and other business activities, although increasing use of information technology is making business clustering of this type less necessary, and many corporations are now looking at cheaper out-of-town office locations.

Toronto's City and Metro Halls and the Ontario Legislative Assembly buildings are in the CBD as well as most provincial government offices, foreign consulates and overseas trade missions.

The CBD is the hub of many transport routes, all of the subway lines converge on the CBD which is also the centre

of a huge regional transportation system run by the Government of Ontario: the main bus terminal and Union Station are located towards the south of the CBD. Most of the office buildings are linked by underground pedestrian routes called PATH (Figure 3.16). Although these are very useful during the cold winter months, the system was not planned coherently and routes between the 1100 stores connected by the system can be confusing.

Figure 3.15   The CBD of Toronto.

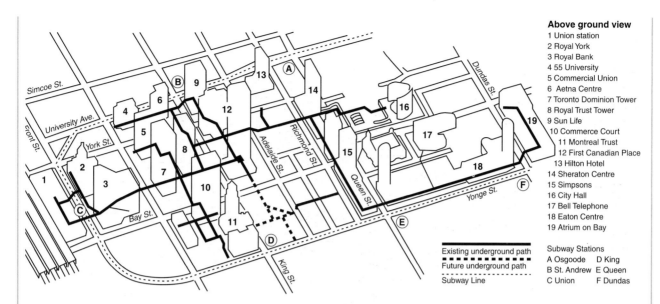

Above ground view
1 Union station
2 Royal York
3 Royal Bank
4 55 University
5 Commercial Union
6 Aetna Centre
7 Toronto Dominion Tower
8 Royal Trust Tower
9 Sun Life
10 Commerce Court
11 Montreal Trust
12 First Canadian Place
13 Hilton Hotel
14 Sheraton Centre
15 Simpsons
16 City Hall
17 Bell Telephone
18 Eaton Centre
19 Atrium on Bay

Existing underground path
Future underground path
Subway Line

Subway Stations
A Osgoode   D King
B St. Andrew E Queen
C Union   F Dundas

Figure 3.16   Underground pathways and malls in Toronto's CBD.

Despite the rise of the suburban shopping mall, many people prefer the noise and activity of Yonge Street (Figure 3.14) for shopping which has thousands of stores, and also the Eaton Centre, a CDB shopping mall, easily reached by subway link from all parts of the city. The centre has 300 shops, cinemas and restaurants on four climate-controlled levels which span many city blocks. This downtown core retailing area ranks third in retail sales in North America, after Manhattan and Chicago. In 1998, 17 600 retail businesses provided over 11% of the city's employment.

Finally, Toronto's CBD is also a tourist and entertainment centre boasting attractions such as the CN tower, theatres, concert halls and sports arenas e.g. the SkyDome Stadium with its unique retracting roof (Figure 3.17).

Retailing

About 5% of land use in the Greater Toronto Area is devoted to commerce i.e. shops and offices. Despite moves towards **decentralisation** of both retail and commerce from the CBD, the greatest concentration of both remains in the city centre. Outside the city centre it is possible to identify three different levels of provision:

- **Strip-commercial** – small stores, offices, petrol stations lining both sides of a major street in a ribbon-pattern. Arteries such as Yonge Street and Bloor Street (Figure 3.14) are examples.

- **Shopping plazas** – usually consisting of around 10-20 shops of a low-order type serving neighbourhood areas of a few thousand people. They usually include a food store, variety store, bakery, drug store and a few other small outlets.

- **Regional shopping malls** – there are currently nine of these huge enclosed centres in Toronto, strategically located near major highways with massive free parking lots. Each is anchored by one or more major Canadian department store along with 100 to 200 chain and speciality stores.

Suburban Business Districts: the emergence of Edge Cities

Although the population of Toronto is gradually moving away from the older parts of the city, most employment in offices and retailing remains in the CBD.

Figure 3.17   The SkyDome and CN Tower.

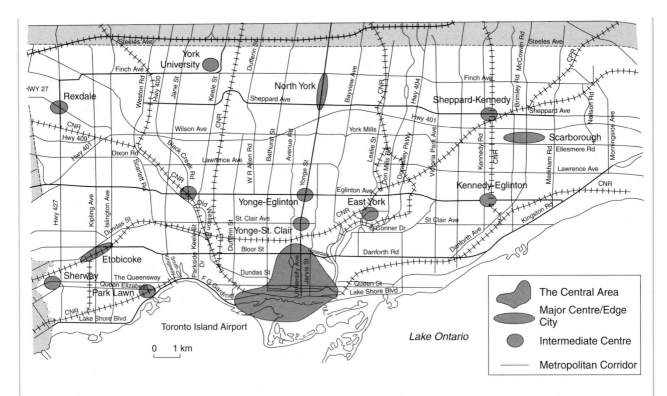

Figure 3.18　Metropolitan centres and corridors in Toronto.

The central area retains around 500 000 jobs, half of which are held by people commuting from outwith the City of Toronto each day. Overall land use figures indicate a slight increase in office use but hide the massive expansion which has taken place mainly through **vertical development** in recent years due to high land values. Since 1971 the total amount of office space in central Toronto has more than tripled. This increase, together with the population movement to the suburbs, has resulted in greatly increased pressure on the highway and public transportation system. As with retailing, there has been a trend towards the development of **suburban business districts** notably in North York, Scarborough and Etobicoke (Figure 3.18). All are located on subway lines and that at Scarborough is at the end of the new light rapid transit system and combines the Education Centre with several high rise office blocks and a huge shopping mall. By 2000 the complex was employing 40 000 people.

The development of certain growth points on the edges of existing cities, which may consist of business parks, shopping complexes and new housing areas has been identified as a major new urban phenomenon – the **edge city.** In the City of Toronto, it is possible to identify a number of these, e.g. North York, Scarborough and Mississauga.

Although they are at different stages of development, these growth points have certain common characteristics:

- more than 1.5 million m² of leasable office space
- more than 180 000 m² of leasable retail space
- more jobs than bedrooms
- most growth has taken place within the last 30 years
- is perceived by the public as one place.

Significantly, these edge cities now mirror the downtown functions of the CBD. In North York, cheaper land and rents have attracted several corporations to move their headquarters or other operations to the new development. More than 3.2 million m² of space is provided with the creation of 100 000 new jobs in shops, hotels and offices. A total of 60 000 people are expected to live in the new highrise flats and **condominiums**. The scheme has been criticised on several fronts – for lacking human scale, being sterile and uninviting, and for the fact that none of the new housing will be aimed at low-income families.

Questions

19 Outline how and why urban growth has affected local government in the Toronto area since 1950.

20 Summarise the main land uses within Toronto's CBD.

21 Identify four different levels of retail provision in Toronto and comment on their functions.

22 Explain the reasons behind the emergence of Suburban Business Districts.

23 Identify examples of Edge Cities in Toronto's urban structure and outline their main characteristics.

Industry

With more than 4 million residents and 2.5 million jobs, this is the most populated and largest economic region in Canada. Toronto has the largest labour force of any city in Canada. One in five employees works in manufacturing or warehousing, and the Toronto region has over 7000 manufacturing companies accounting for 21% of the Canadian total. Traditionally, this area has been the industrial heartland of the country. The late twentieth century saw considerable industrial change and restructuring but car-making, telecommunications, pharmaceuticals, chemicals and plastics and the aerospace industries continue to expand and invest in the Toronto area. The importance of this mix of traditional and emerging industries lies in their potential for future growth. In particular the area's positive economic outlook is enhanced by:

- easy access to large markets in the USA
- a centrally located international airport
- a highly integrated motorway system
- a well educated labour force
- training, research and development at local universities.

When considering the industrial geography of the Toronto area, it is important to recognise the economic forces and the historical factors which have resulted in the existing pattern of industrial activity. Any analysis of this distribution requires an understanding of the many influences which provide certain locations with competitive advantages.

Before 1850 manufacturing was largely confined to small workshops within the city area. Small markets for manufactured goods and limited accessibility to those markets were the main factors responsible for the small scale of production. After 1850 the emergence of the railway led to a significant improvement in accessibility to markets, and towards the end of the nineteenth century Toronto emerged as a centre of industrial activity. This continued into the twentieth century, with the advent of mass production technologies and assembly lines which contributed to single-storey modern factories. Increasingly larger scales of production in these factories began to have significant negative impacts on the areas immediately around, and there was a need to separate industrial activities from other land uses. Additionally, with increased dependence on road and air transport, new industrial parks characterised by large, low density, single-purpose buildings began to emerge near to the international airport and the major highways around Toronto (Figure 3.13).

Distribution of Industry

Between 10 and 15% of Toronto's land is used for industry. The distribution of manufacturing industry (Figure 3.13) reflects the continuing importance of historical factors and inertia in the inner city area, and the outstanding importance of road transport in the newer areas of the outer suburbs.

Western and eastern edges of the CBD

These areas became important in the late nineteenth and early twentieth centuries, having access to both labour and markets. They are now in decline due to outdated premises and the difficulties of expansion. The main industry still located in this area is clothing which employs 10 000 workers (often female immigrant workers of Greek, Italian and Portuguese origin). The area acts as a 'conveyor belt' with small companies each specialising in individual stages of making a garment. Wholesalers and distributors are also concentrated in this area and high land prices can be borne by having firms on different floors of multistorey buildings.

Don/Harbour Area

This is the oldest industrial area, based around the harbour where many industries requiring imported raw materials located to cut costs. Food processing (flour milling, sugar refining, whisky distilling) used raw materials brought in by water, and used the lake for waste disposal. The main railway yards were adjacent, and the area also developed meat processing factories as cattle were brought into the city by rail from other parts of the country.

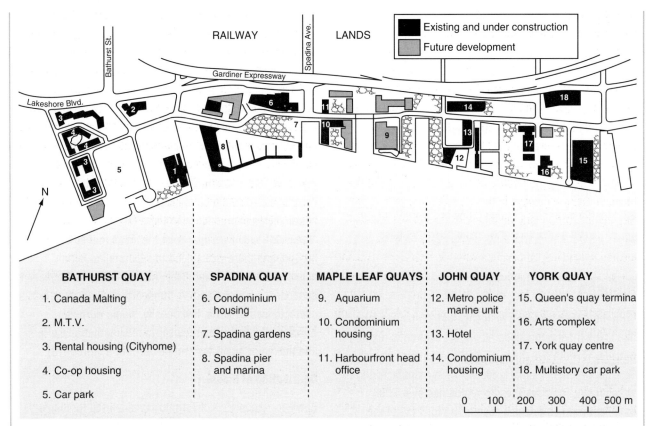

BATHURST QUAY	SPADINA QUAY	MAPLE LEAF QUAYS	JOHN QUAY	YORK QUAY
1. Canada Malting	6. Condominium housing	9. Aquarium	12. Metro police marine unit	15. Queen's quay termina
2. M.T.V.	7. Spadina gardens	10. Condominium housing	13. Hotel	16. Arts complex
3. Rental housing (Cityhome)	8. Spadina pier and marina	11. Harbourfront head office	14. Condominium housing	17. York quay centre
4. Co-op housing				18. Multistory car park
5. Car park				

0   100   200   300   400   500 m

Figure 3.19   The Harbourfront area.

The area was flat, easy to build on and was close to the workers' houses at a time when there was little personal transport. Today this area is undergoing a major change; no space for expansion, congested roads and no parking are forcing industries to relocate. As they do so, the area is gradually being transformed into housing and recreation space.

Harbourfront

Although Toronto is a lakeshore city, until recently access to Lake Ontario has been restricted by the railway tracks which presented a physical barrier to integrating the CBD with the lake shore area. The last few years have seen the complete transformation of the old harbourfront area as new commercial, entertainment and residential developers have seen its potential (Figure 3.19). The area by the lake was previously an industrial and dockland zone of warehouses and factories. These are now rapidly being turned into art and craft galleries, shops and restaurants, marinas and **condominiums** (Figure 3.20). The last is vital because it is an important part of the plan to have people living full-time in the area (although so many highrise apartment buildings have now been constructed that some people argue that they have created a new barrier between the city and its waterfront).

Between Harbourfront and the CBD other new developments are under construction in a large area formerly used for railway yards for assembling freight trains. The area is adjacent to the CBD and therefore a prime development site. The CN Tower (at 533 m, the world's largest free-standing structure) and the SkyDome, with its retractable roof are both located here (Figure 3.17). A programme of further development began in 1993 and recently new office buildings, shopping and entertainment areas and apartments for 15 000 people have been completed. It is the largest redevelopment project in Toronto's history and is scheduled for completion by 2010.

Figure 3.20   Harbourfront, Toronto.

Industrial parks in the suburbs

Land and taxes are cheaper in the regions outside the central area and many companies have relocated to sites which allow them to build modern space-consuming single-storey buildings, suited to modern automated technology. Access to highways is another major factor, as is the avoidance of older, congested sites. The labour supply has also relocated to the suburbs. One of the major concentrations of industrial parks is near the international airport, reflecting the importance of air transport and the focus of major highways in the area. One example is the Skyway Industrial Park, built near highways 401 and 427, where modern industrial and commercial premises are located in landscaped grounds with wide service roads and car parks. The type of industries attracted to the area are of the 'light-secondary' tertiary/quaternary category i.e. data processing, insurance offices, warehousing, packaging and light assembly work.

Deindustrialisation and relocation

In planning industrial strategies for the twenty-first century, planning authorities in Toronto have identified the relatively high percentage of vacant industrial floor space in the central area of the city as '**deindustrialisation**' of these areas. City planners are now concerned over the loss of manufacturing jobs to the outer edges of the built-up area such as North York, Scarborough and Etobicoke and beyond into the regions. Between 1950 and 2000 the number of industrial jobs in the central area halved. Although increased office jobs have offset this loss to some degree, the economic base of the central area has been narrowed, and job opportunities reduced. Massey-Ferguson was one of the largest inner-city firms, but it contracted from 2000 workers in 1976 to 500 workers in 1985. Despite reducing the size of the plant, overheads remained high and traffic congestion reduced efficiency. The entire site has now been closed and has been developed for housing and some light industry. Massey-Ferguson have relocated to a cheaper site at Blantford between Toronto and London, Ontario. The Toronto Economic Development Corporation was created to maintain a diversified economy in the centre, but despite incentives for prospective employers, high land prices, property taxes and traffic congestion continue to force expanding manufacturers to re-locate in the regions which border the city. This reveals a clear pattern of **industrial suburbanisation.**

The effect of this deindustrialisation on the ground is very clear to see. Demolition of older industrial buildings and the resulting vacant land and possible soil contamination discourage potential developers from locating in these areas and encourage further decline. Once thriving concentrations of employment within the city are becoming areas of derelict land. In contrast, as more and more land on the fringes of the built-up area is converted to industrial use, farmland is lost to developers and the boundaries of the urbanised area expand ever outwards. Perhaps the most significant implication of the changing distribution of industrial activity in Toronto is its effect on the efficient use of the urban infrastructure. The abandonment of former industrial land in the central area leaves existing services such as utilities and transport unused. At the same time, the level of demand for these services in high growth areas on the urban fringe increases.

Questions

24 What advantages does Toronto have as an industrial location?

25 Explain what you understand by the term 'industrial suburbanisation'.

26 What problems have been caused by industrial suburbanisation in Toronto for:
  a inner city areas
  b peripheral areas of the city?

27 Identify and explain why some inner city industries have not relocated to the suburbs.

28 Outline the redevelopment of the Harbourfront area of Toronto.

Housing

Over 80% of land in Toronto is zoned for residential use. As shown in Figure 3.13, most of the land used for housing is in the suburbs. However, Toronto is unusual amongst major North American cities in having a considerable number of residential buildings in or near the downtown area. Most of these are highrise apartment buildings and obtaining low cost housing is very difficult. At the end of the twentieth century new condominiums were still being built in the CBD. 'Condos' are simply apartments which are owned by their residents. This unusual concentration of residents in central areas means that Toronto's CBD is busy almost 24 hours a day, helping to create a safer downtown.

Figure 3.21   Detached housing in the suburbs of Toronto.

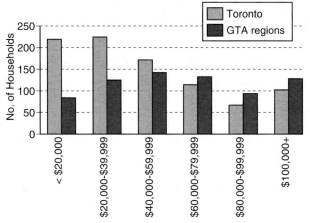

Figure 3.22   Households by income groups, 1996.

This is reflected in the title of the city's 1990s structure plan, '**The Liveable Metropolis**'.

Within Toronto the choice of residence is dictated largely by income. The cost of housing is increasing faster than in any other Canadian city. For most residents, the preferred choice of accommodation is a detached house in the suburbs (Figure 3.21), but for many this is becoming less possible and **income polarisation** is increasing. This means that most households in the city are concentrated at the lower end of the income spectrum, whilst those in the GTA regions are polarised at the high end. In 1966 the City of Toronto had 70% of the area's lower income households, whilst the surrounding regions had 56% of all high income households (Figure 3.22). Such polarisation between the **core** and **periphery** is common and perhaps even more pronounced than in other cities in North America. Such an imbalance creates problems for Toronto's tax base and provision of social and human services. It is important to realise that just what makes up the 'core' and the 'periphery' is subject to change through time. Older housing predominates in the downtown area, with postwar housing in the outer city areas of Etobicoke, North York and Scarborough. These areas were the periphery in the middle of the twentieth century and grew rapidly after 1945 as young families moved out to suburban neighbourhoods where identical houses were built over vast areas. Scarborough in particular became the subject of many jokes as 'Scarberia' or the 'bedroom city' dormitory suburb.

During the 1970s new spatial patterns in housing emerged with huge increases in population in the fringe regions around the city (which were the new peripheral areas as suburbanisation had spread even further from downtown).

Between 1971 and 1981 Etobicoke, North York and Scarborough lost population as some of their residents moved even further from the CBD to the newly developing outer suburbs.

Today in the city itself there is a concentration of rental apartments, condominiums and assisted housing. Low to moderate income families live here, together with single people, immigrants and senior citizens. In the outer regions, detached housing predominates and this region is predominantly populated by young, upper income families.

It is almost exclusively young families buying a home in the outer regions, where their preferred house type (a detached house) is more affordable, who leave the city. During the 1990s each year over 100 000 people left the city to live in the surrounding regions. Net migration from the city accounted for half of the GTA regions' growth during this decade.

Immigration accounts for a significant flow into the city each year, over 70% of whom settle in the city because of the availability of rented accommodation. In addition, there are return flows of migrants from the GTA regions, these are mostly students leaving the family suburban home and renting flats and rooms in the inner city near the universities.

Since the flow of population out of the city is overwhelmingly the young, there is a knock-on effect on the structure of the population which remains in the city, which is only marginally tempered by immigration. By 2011 over half of the city's homeowners will be over 55, and a similar ageing pattern is evident amongst the tenant population, many of whom are unable to afford the move to the higher cost housing in the suburbs.

Inner city redevelopment: St Lawrence

A number of major inner city renewal projects were undertaken by the city towards the end of the twentieth century. They were part of a process initiated by the city planners called **reurbanisation** which is intended to reverse the seemingly unstoppable outward urban sprawl by undertaking **urban renewal** projects in areas where the **urban fabric** of services is already in place.

The St Lawrence area (Figure 3.14) is located just to the east of the CBD and was the largest inner city housing redevelopment scheme undertaken during the twentieth century. The area is spatially distinct with the office and retailing core on one side, an old industrial area to the east, and the railway to the south. Working class housing was developed in the St Lawrence area during the nineteenth century along with storage and distribution facilities for industry. By the mid 1970s, housing quality had declined and the industrial functions had largely moved out to the suburbs leaving large areas derelict. The city council purchased any land it did not already own and established a plan for the area based on integrating the new development with the surrounding area and avoiding becoming a public housing ghetto for poorer income families:

- traditional street patterns were largely retained with shops and medium rise apartment blocks

- several architects were employed to avoid the 'Scarberia' factor

- a mix of private and subsidised housing.

As shown in Figure 3.23 the redevelopment included a new school, a young people's theatre, and retained the famous St Lawrence Market, a popular local indoor market much used by CBD office workers.

The development of the St Lawrence area was seen as a rejection of the policies of city redevelopment, which are still popular in many European cities, where high rise flats were thought to provide the answer to inner city housing problems. It also avoided the pitfalls of inner city redevelopments in USA, notably in Detroit and Chicago, which became no-go terror zones. The scheme has been completed and now houses some 10 000 people with a mix of household types, incomes and lifestyles. The success of the St Lawrence neighbourhood has proved that Toronto can deliver on its promise of a more liveable city.

Open spaces

From the Toronto Structure Plan:
*'The Metropolitan open space system shall continue to be based primarily on the major **river valleys** and the **waterfront** ...'*

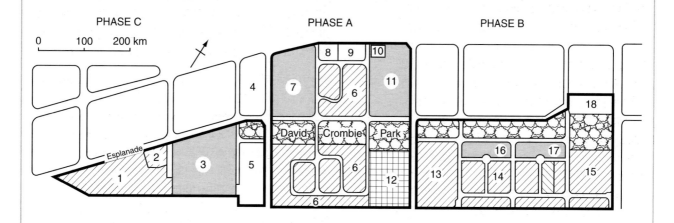

1. Private housing
2. Private Condomimium Units & Commercial
3. Mixed Private/Public Housing
4. St. Lawrence Market
5. Gross Machinery Site
6. Housing
7. Market Site, Housing & Commercial
8. Private Condomimium Units
9. Private Commercial Development
10. Young People's Theatre
11. Cityhome
12. Transformer Site
13. Cityhome
14. Private Housing
15. Co-op Housing
16. Private Commercial Development
17. Co-op Housing
18. School Site

Figure 3.23   Redevelopment of the St Lawrence area.

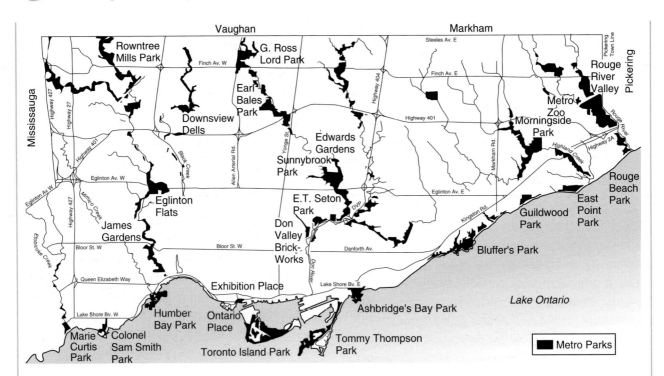

Figure 3.24   Open spaces in Toronto.

Despite the rapid growth of the city since 1950, Toronto has managed to retain many unique and interesting natural areas. Currently 3273 hectares of land are controlled by the Metro Parks department and include managed areas such as horticultural and zoological gardens and golf courses, and wilderness areas deliberately left as natural as possible (Figure 3.24). The city's many natural features are mainly found along the Lake Ontario waterfront and along the river valley and stream corridors. These natural areas include an abundance of unique geological and biological features which support equally unique and diverse plant and animal life and associated habitats. Toronto is a remarkably green city with its natural areas offering a range of outdoor recreational opportunities.

The Valley Lands

The Humber, Don and Rouge Rivers, together with several tributary creeks have cut deeply incised valleys through the easily eroded sediments of the former lake (Iroquois) bed. These slopes are relatively unstable and building regulations prohibit construction on them. In many cases they have been afforested to help with **slope stabilisation** and reduce runoff. This has also enhanced their scenic attractiveness as they cross the cityscape. The shape of the open spaces follows the valleys (Figure 3.24) and they are ideal for walkways and cycletracks.

The large block of open space in the east (Figure 3.24) is the Rouge river valley which has been designated as Canada's largest urban park (4250 hectares). The land has remained undisturbed because it was purchased in the mid 1970s as greenbelt land between the city and a proposed new airport. Once the airport proposal was dropped, the fight over the Rouge valley began between developers and conservationists. The developers rented space on billboards next to the expressways and commuter routes near the valley and put up adverts claiming, *'If it weren't for a few deer, you'd be home by now!'* The conservationists pointed out the effects of development near the mouth of the river where manicured park lawns have replaced the tall grasses and other natural bird habitats. Further west, the Don and Humber Valleys are even worse. Little is left of the Don's marshes and the Humber's are stagnant with toxic industrial effluent. In contrast, the Rouge valley remains mostly unspoilt with oak, cedar hemlock and maple woodlands and open farmland. Even with park protection, pressure from the surrounding urban area still exists e.g. to use the valley for landfill sites, golf courses and even cemeteries! The valley represents one of the Toronto's few remaining intact ecosystems and will need the protection given by its recent designation as an urban park.

The Waterfront

Toronto's lake frontage gives it great potential for a range of watersports and shoreline recreational activities. One project already very successful in this respect is the Harbourfront development discussed on page 118. Other schemes include the Toronto Islands with walks, beaches and a traditional Canadian village; and Ontario Place, a 40 hectare recreation complex including a marina, an arena for rock concerts and a six storey Imax cinema. An important part of the city's plan for this area is to create a lakeshore walk/cycleway linking the Ontario Place and Harbourfront areas.

Transport

The shift in population distribution towards the regions around the City of Toronto has had a number of implications for city planners. Perhaps the most obvious problem has resulted from increased **commuting** from these outer suburbs into the city, particularly to the CBD where most jobs are still concentrated despite moves towards decentralisation. The population of the outer regions are heavily dependent on the car with most families having two or three vehicles to cater for their needs. Services in the suburbs, particularly shops, schools and health care have become road-orientated, and this further encourages **urban sprawl.**

Road Transport

There are more than 1.5 million cars in Toronto, and a further 1.5 million are driven into the city every weekday from the surrounding regions. This puts considerable pressure on the city's road system and traffic congestion is increasing. An extensive expressway system was constructed in the late 1960s which had a major detrimental impact on the environment due to noise and air pollution. During the course of road construction, the city lost a large number of old buildings which in some cases were historically valuable. The road congestion is exacerbated because there have been no new motorways constructed in the central area of Toronto since 1971, when the proposed Spadina Expressway plan which would have bisected the CBD was rejected after widespread public opposition. There has been a major anti-road lobby in the city for over 30 years with many people arguing that money should be spent on improving public transport.

Once car drivers have arrived in the city, they face the further problem of trying to find a parking space. Although some city centre offices provide underground car parking for some employees, this is usually only available for a few top executives and the costs for other car-borne workers can be very high.

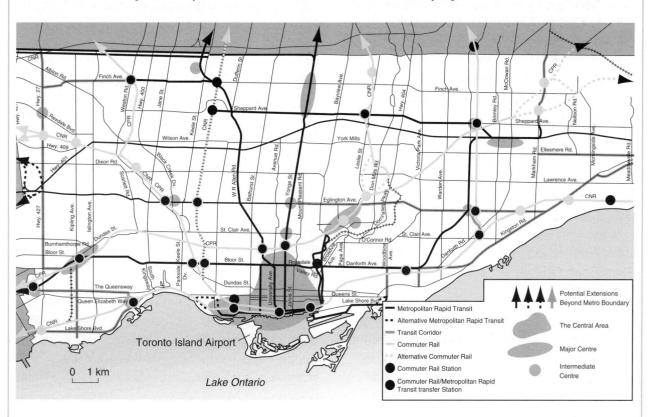

Figure 3.25   Transport systems in Toronto.

Parking charges in the CBD are very high due to the high land value. Some commuters have contracted spaces with commercial car parking companies which guarantees a space for a high annual or monthly fee.

Public Transport

The Toronto planning authorities have adopted a policy of encouraging public transport at the expense of the private motorist in a deliberate attempt to combat the pressure of commuter road traffic. This policy is in sharp contrast to most cities in North America where road investment has always taken priority over public transport. In Toronto the urban transit lines form a vast network of routes linking the city's suburbs to its CBD carrying millions of commuters in and out of the city each day (Figure 3.25). The system is administered by the **Toronto Transit Commission** (TTC) which operates buses, trolley buses, street cars and subways in a fully **integrated transport network** of over 4000 miles of route. Current usership figures are approaching 400 million passenger trips each year. Figure 3.25 illustrates the important link which has developed between the public transport system and the move towards out of town office centres such as those at Etobicoke, North York and Scarborough.

The policy of the TTC is to provide public transport to within 600 metres of 95% of Toronto's population. Its success has been due to the fully integrated nature of the system e.g. journeys which involve a change of type of transport do not involve additional charges and the interchanges are designed to minimise waiting and inconvenience (Figure 3.26).

Land values increase greatly when public transport is improved, for example in the 40 years since the opening of the Yonge Street subway line, 50% of all new apartment construction in that part of Toronto was built within walking distance of the line. During the same period 90% of all new office construction took place near the major subway stations, and all of the new out-of-town centres at Scarborough, Etobicoke and North York have rapid-transit stations.

The development of the **GO train** (Figure 3.27) network which reaches out into the developing regions outside the city now handles over 37 million passengers each year from 44 local stations. Union Station is the hub of the GO network where it links directly into the TTC subway system.

Figure 3.26　A transit link in Toronto.

The future

A recent planning report forecast an increase of 55% in the number of daily transport trips in Toronto by 2021 and since neither the existing TTC network, nor the road system could cope with this, the city is planning major transport developments on all fronts. Future plans include new subway lines and a **Light Rapid Transit** (LRT) line has recently opened in the harbour area connecting Union Station with Harbourfront, the CN Tower and the SkyDome. This LRT is intended to encourage suburban dwellers to use public transport to reach the SkyDome for matches. The road traffic chaos in the area before and after major events is a major factor in encouraging the use of public transport.

Infrastructure water and waste

As population has grown in the Toronto area, the demands on services such as water supply and waste disposal have intensified. The people of Toronto use vast quantities of water every day for domestic and industrial purposes, but

Figure 3.27　GO trains in Toronto.

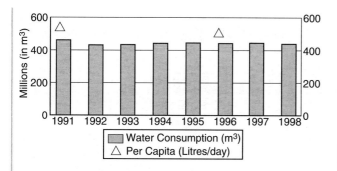

Figure 3.28   City of Toronto water consumption, 1991–1998.

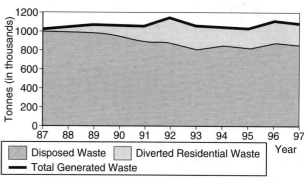

Figure 3.29   Municipal solid waste, City of Toronto, 1987–1997.

within the last decade city residents have become more conservation conscious as the need for sustainable use of resources has been recognised through education programmes and advertisements. These have contributed to a decrease in residential per capita water use (Figure 3.28), although consumption across the city has increased due to population growth and improved economic activity.

The most successful conservation programme in the city relates to solid waste (Figure 3.29). During the late 1980s a scheme was developed to divert waste from landfill sites. Since then diverted waste increased by over 1000% from 21 553 tonnes in 1987 to almost 250 000 tonnes in 1998. The most innovative component of the waste diversion scheme is the city's '**Blue Box System**' whereby residents pre-sort rubbish before collection into separate boxes for paper, metal, plastics etc.

Managing Urban Change

The major successes in managing urban change in Toronto are in part due to the changes in local government organisation. There can be no question that the Greater Toronto Area is a single urban entity, marked by an older core city, surrounded by newer regional areas. As the city has expanded, there has been recognition that planning for housing, employment, transport and the environment can only be undertaken for the entire GTA area. At the same time it has been recognised that there are differences between Toronto and the GTA regions, rooted mainly in the city's complex socio-demographic mix and its mature urban form. As we have seen, Toronto still faces many challenges; social, economic and environmental if it is to retain its claim to be 'the liveable metropolis'.

Questions

29 Explain the link between income polarisation and the social structure of the GTA area.

30 Describe the movement of population within and between the City of Toronto and the surrounding regions.

31 In what ways does the St Lawrence area of Toronto illustrate the city's 'liveable metropolis' claim?

32 Summarise the main elements of Toronto's open spaces system and outline how they may be under pressure.

33 What problems face drivers in Toronto?

34 Explain how the TTC's Integrated Transport System helps to resolve traffic congestion in the city.

35 What successes have been achieved in managing public services such as water supply and waste disposal in Toronto?

Urbanisation in Pakistan

Understanding the current urban processes at work in an area really requires a look back at the past. This applies equally to developed and developing countries, even though the timescales of **urbanisation** may vary enormously.

The present-day boundaries of Pakistan (Figure 3.30) contain the locations of the oldest urban settlements in the Indian sub-continent (Figure 3.31) of which Harappa and Mohenjo Daro are the best known.

These civilisations, based on the agricultural potential of the Indus basin, date back to 3000 BC, with urban development at its peak from 2000 BC to 1700 BC.

Phases of development

The main phases in the development of the modern cities of Pakistan are shown below. In each case, the examples chosen are cities which can trace their origins back to the particular phase under which they are listed.

- **Pre-historic (3000 BC)** Harrappan civilisation has been traced through archaeological findings. There was a network of craft towns and a settled agricultural system based on irrigation. Settlements like Mohenjo Daro are no longer in use but the original site of Lahore dates from this era.

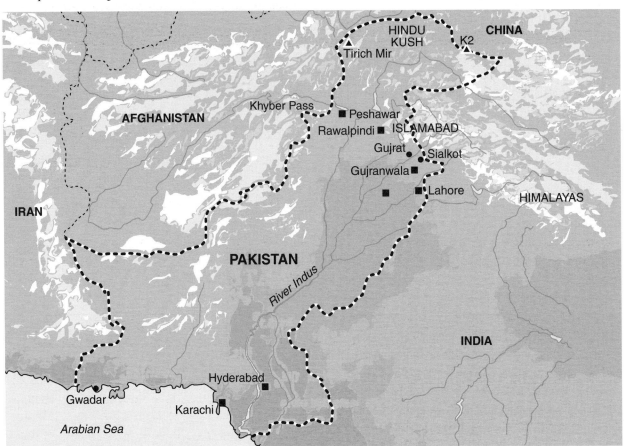

Figure 3.30   Satellite image of Pakistan.

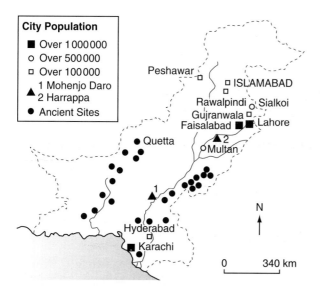

Figure 3.31   Urban centres in Pakistan.

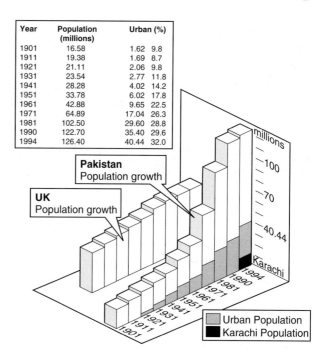

Year	Population (millions)	Urban	(%)
1901	16.58	1.62	9.8
1911	19.38	1.69	8.7
1921	21.11	2.06	9.8
1931	23.54	2.77	11.8
1941	28.28	4.02	14.2
1951	33.78	6.02	17.8
1961	42.88	9.65	22.5
1971	64.89	17.04	26.3
1981	102.50	29.60	28.8
1990	122.70	35.40	29.6
1994	126.40	40.44	32.0

Figure 3.32   Population growth in Pakistan, 1901–1994.

- **Kingdoms (AD 1000–AD 1200)** Cities like Qureshi and Lahore grew up as trade, religious and administrative centres for early Buddhist and Hindu kingdoms. Many of these early cities were partially destroyed by subsequent rulers and then saw a rebirth in their fortunes 200 years later in the Muslim period.
- **Muslim Era (AD 1500–AD 1800)** Many of the cities of the early kingdoms dropped out of historical sight as they were attacked, devastated and abandoned. The Mughal Empire conquered the Punjab in the early part of the sixteenth century and this era saw a rebirth of the older settlements and the growth of newer towns. Quetta and Peshawar date back to this period.
- **Colonial (1840–1947)** Under the domination of the **British Raj** there was an expansion of settlements linked to the development of ports, railways and the military. Karachi and Hyderabad are the main cities from this era.
- **Post-colonial (1947–present)** Urban development was linked to the production of five-year plans, a legacy of the bureaucracy from the colonial period. The new capital of Islamabad was constructed as an exercise in town planning.

Urban population growth in Pakistan

It is vital to set the urbanisation process against the background of the overall growth of population in the country (Figure 3.32).

The first full census of India (which at that time included present-day Pakistan) took place in 1871. Wherever possible, the figures for Pakistan are projected back to allow a clearer picture to emerge. One of the most significant trends noted from a study of these figures is that Pakistan is one of the few countries in Asia with a population growth rate as high as 3% per annum.

From Figure 3.32 it is also clear that while the urban population of Pakistan is only around 30% of the total population at present, the growth rate of this urban population is running at a higher rate than the overall growth rate for the country, thus providing statistical evidence for the process of increasing urbanisation. While the evidence of increasing urbanisation is clear, it is more important to try to understand the causes of the process, and to look at the results in terms of the effects on the people and their country.

The root causes of the urbanisation process lie predominantly in the rural areas, where the 'push' factors drive increasing numbers of people, both seasonally and permanently, in search of employment and accommodation in the cities. Many of Karachi's immigrants are in fact itinerant agricultural workers who come into the city during those times of the year when they cannot find employment on farms in the surrounding area. Eventually, and especially during prolonged periods of drought, many will settle permanently in the city, thus swelling the numbers in the **katchi-abadis (squatter settlements)**. In order to stem the flow of migrants, it is necessary for the government to address the problems faced by farmers in the rural **hinterland**, and to seek to establish alternative means of employment for them in suitable industries.

The results of increasing urbanisation are clearly evident in the cities themselves, where perpetual change and the burgeoning population complicate an already difficult management situation for the politicians and planners. Despite the problems faced by many cities in the developing world they continue to exert a powerful attraction for those seeking a way out of the poverty and misery of the rural areas, so clearly many 'pull' factors continue to operate. These include the potential for employment and the expectation of accommodation, both of which all too often fail to materialise.

Urban trends in Pakistan

The smaller urban centres in the northern regions of Peshawar and Rawalpindi (Figure 3.31) show the greatest rates of recent growth. This is due to the magnet of Islamabad, which is a new component in the urban system in Pakistan. This growth reflects the growing economic importance of the north, and the current relative decline in the importance of Karachi and south west Pakistan, indicative of a long term shift in the political geography of the state.

Urbanisation in Pakistan can be summarised at two levels:

Regional 'push' and 'pull' factors operate in and around the major cities and influence rates of growth.

Local (city level) – the character of urbanisation is shaped by the incoming population. Its requirements shape urban planning policies, and these change rapidly as the character of the population alters.

As is shown in the following case study, the planners in Karachi have been forced to accept the continued existence of squatter settlements, to manage their improvement and control their development, rather than try to remove them altogether (Figure 3.33). In this experience, Karachi exhibits characteristics which are very similar to those in many other developing cities.

Questions

1 Compare the figures for Pakistan's total population and urban population over the ten-year period shown in Table 3.5, and briefly try to account for the differences between them.

	Total Population (millions)	**% growth**	**Urban Population (million)**	**% growth**
1981	100	3.0	30	5.0
1991	110	3.1	40	7.0

SOURCE: ESTIMATES BASED ON UN POPULATION YEARBOOK DATA

Table 3.5 Pakistan's urban growth.

2 From the figures provided in Table 3.6, construct a rank order table for each statistic.

	Population (million)	**% growth**	**% urban**
India	911.6	1.9	26
Pakistan	126.4	3.0	28
Bangladesh	116.6	2.4	14
Sri Lanka	17.9	1.5	22
Afghanistan	17.8	2.8	18

SOURCE: ESTIMATES BASED ON UN POPULATION YEARBOOK DATA

Table 3.6 Urban population in selected countries in Asia, 1994.

3 Look at Table 3.7. It gives two different predictions for Pakistan's future population growth. Which prediction do you think is likely to occur? Give reasons for your answer.

	Total Population (million)	% growth	Urban Population (million)	% growth
A	125	3.0	55	8.0
B	115	2.5	50	7.0

SOURCE: MASTERPLAN 2000 – KARACHI DEVELOPMENT AUTHORITY

Table 3.7 Predictions for the population of Pakistan in 2001.

Figure 3.33 Karachi skyline.

Case Study Karachi

Site, situation and structure

Karachi is situated on the shores of the Arabian Sea, to the north of the mouth of the River Indus (Figure 3.34). It is a relatively young city; there was only a small fishing village at the mouth of the Lyari River, and a fortification at Manora right up until the early part of the nineteenth century.

In 1839, the fort was taken by the British, who then began the development of the modern city, inland from the original coastal village. They dredged and improved the natural harbour, and laid out a street plan in the area between the Lyari and Malir rivers (Figure 3.35), thus beginning the landward expansion of the city which has continued to the present day.

Figure 3.34 The location of Karachi.

Partly due to its strategic importance, and also because it was cooler than the original capital at Hyderabad, Karachi became the capital of Sind province in 1843. With the expansion of the railway system in the 1860s and continued improvements to the now bustling harbour, its commercial and industrial future was assured. This was despite the economic dominance held by Bombay throughout this period, right up to **Partition** and the creation of the state of Pakistan in 1947.

When the new state was created, the capital was established in Karachi (though this was to move to Islamabad in 1971). However, the initial impetus this gave to the commercial and industrial life of the city ensured that Karachi quickly became the largest and most important city in Pakistan, with the largest port, a major international airport, main rail and road termini and a wide variety of industries.

The **Metropolitan Area** now houses 6% of Pakistan's total population, and 22% of its urban population. It generates 15% of the nation's Gross Domestic Product (GDP), and accounts for 35% of employment in large scale manufacturing.

Karachi's rapid population growth can be seen in Figure 3.36. Most of the current growth stems from immigrants from the surrounding agricultural hinterland, or from farther afield in Pakistan. Many immigrants previously arrived in the area as refugees from India, as a result of the partition of the two countries which coincided with Independence in 1947.

129

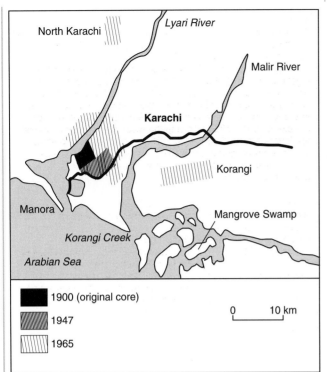

Figure 3.35   The early growth of Karachi.

1947
By independence in 1947 considerable expansion had occurred to the east of the original centre. This was linked to port development and the growth of the railways connecting Karachi with the rest of the Indian sub-continent. The provision of paved roads and streets lagged far behind, and played little part in the direction of physical growth of the town.

1965
By 1965 improvements to the road network, though patchy and often poorly planned, had begun to be felt, and the city had expanded in all directions.

Two main areas of expansion can be noted.
1 The townships of Korangi and Landhi (both established next to new industrial areas) far to the east of the centre
2 North Karachi
These were early attempts to remove people from some of the slums in the city and provide them with better houses, services and jobs about 17 km from the centre. Unfortunately most who moved there were forced to continue working in the central area, and many tried to return to houses nearer their jobs rather than commute this distance.

As a result, Karachi now has a very mixed population; the Sindis, natives of the region, account for less than 10% of the population; while Pathans, Punjabis, Muhajirs and Baluchis together constitute over 90%.

By the time of the 1991 census, the population of Karachi had risen to over 8 million (Figure 3.36), and estimates for the year 2000 are put at over 11 million people. By the year 2015, the United Nations has predicted that this figure could rise to over 20 million.

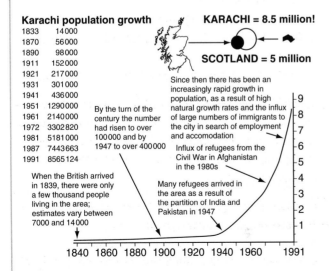

Figure 3.36   Population growth in Karachi, 1840–1991.

The physical growth of Karachi

The present city of Karachi covers an area of about 150 000 hectares (Figure 3.37). In 1947, the total built-up area was only 18 600 hectares, and most of that was due to the rapid expansion under the direction of the British colonial administration in the late nineteeenth century. The original site consisted largely of a group of islands and sandspits. On the landward side were some barren hillocks, and it was between these and along the shore of the main bay that the early fishing villages grew up.

By the time the British arrived a small walled town had developed in what is now the commercial heart of the city, stretching from Kharadar near the shore to Mithadar on the Lyari river. From this centre expansion has taken place mostly to the north and east. Most of the older public buildings date from the late nineteenth century, during the period of British colonial rule.

Many of the housing areas on the fringe of the **CBD** were built to accommodate military personnel and civil servants, and were known as Barracks, Quarters or Lines, names which they retain to this day (Figure 3.38). The Pakistani Army still provides accommodation for its officers in the city in this way.

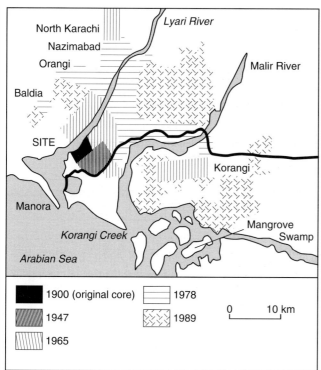

By 1978 further drift had occured, along with a general expansion mostly to the north and east. By this time too, the first developments of the area of Orangi had begun. This is one of the largest of the many Bastis of Karachi, and it consists of a legal area housing about 100 000 people and an illegal area accomodating about 400 000 people. Orangi town and Baldia Colony were the first of the major expansions to the north west of the city: the squatter settlements were pushing out towards a major air base, often following the water courses, since these are ignored by official housing and commercial developments. This has also occured within the city, and accounts for much of the infill of urban growth around what are usually middle-class housing areas.

1989
Since the mid-1970s the growth of housing has been more directly controlled and planned by the authorities and attempts have been made to improve and upgrade the squatter settlements. Much of this has occured under the Karachi Master Plan, which was developed and implemented by the Karachi Development Authority in conjunction with the Karachi Metropolitan Corporation, with help from the United Nations as well as other national agencies and authorities within Pakistan.

Figure 3.37   The modern growth of Karachi.

Managing change in Karachi

The Role of Authorities

Karachi Municipal Corporation (KMC) is the prime local body responsible for municipal affairs of the city. The major functions and responsibilities of KMC include:

- general maintenance of urban areas, including landscape, parks and recreational grounds

- provision of community services such as hospitals, dispensaries, maternity homes, basic health care units, schools, colleges and basic education
- leasing and public property maintenance, i.e. of roads, streets, bridges, markets, public buildings
- construction of roads and flyovers
- co-ordination of planning and development activities
- solid waste management, including sweeping, collection and transporting to landfill sites
- fire fighting services
- squatter site upgrading / management and leasing
- flood control, maintenance of street lights, slaughter houses, grave yards
- sanitation and maintenance of storm water drainage.

Karachi Development Authority (KDA) was formed in 1951 as the Karachi Improvement Trust. In 1957, it was restructured as the Karachi Development Authority. The authority is **autonomous** in nature and performs the following functions:

- master planning, co-ordination, environmental control, urban design and urban renewal
- land management and development control
- maintenance of **infrastructure** in the schemes managed by the development authority.

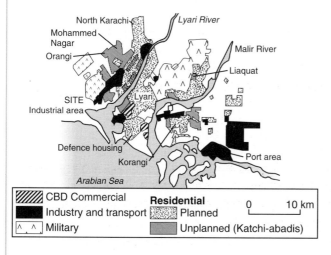

Figure 3.38   Land use patterns in Karachi.

Defence Housing Authority was established in 1953 under the Ministry of Defence. The authority provided residential plots to the retired and serving officers of the armed forces personnel. It now owns the most expensive residential localities of the city and manages an area of around 6500 hectares.

Karachi Port Trust, formed in 1887, is responsible for the maintenance and development of the Karachi port and adjoining ancillary estates. It also provides housing facilities to its employees in the immediate localities.

Planning decisions and the management of change in a city like Karachi pose particular problems for municipal authorities. The city is growing at a faster rate than the politicians and the planners can cope with. The Karachi Development Authority has to work in a reactive way. It can make five-year plans, but because of lack of money and the control to implement the plans they may stay on the drawing board. The problems for the authorities in Karachi are compared with a city in the UK below.

UK	Pakistan
Planning	
In a city in the UK the municipal authority will have a highly-qualified and well-trained planning department. The city will be divided up into defined areas and Local Plans will be drawn up at regular intervals. These provide a framework for control over land use and management of city growth and guidelines for the provision of services and **public utilities**. There is always a period of consultation before a plan is finalised.	In Karachi there are several different bodies which have authority over aspects of town planning. The KDA has the role of making up the actual plans for the city. In the past they have been given help by the United Nations Development Programme (UNDP). The implementation of the plan is divided between several autonomous city agencies with no clear links. The planning and development control system is fragmented with no effective, area-wide system for controlling developments.
Budgeting	
The local authority gains its income from direct local taxation and grants from central government. A huge bureaucracy has been set up to collect these taxes and a yearly planned budget is set to provide services for its citizens.	A large proportion of city-dwellers in Karachi earn low incomes and cannot afford to pay for services which are provided. The bulk of their income will go on food and they cannot afford rent for flats built by the authorities. They are forced to build **shanty dwellings** wherever they can and in many cases these have no access to public services. Theoretical physical planning is not linked to the ability to pay for the actual scheme. In one example, if a plan by the KDA to upgrade slums in one area over a five-year period had actually been carried out, the total cost would have exceeded the total income for the whole Sind region by five times.
Services	
The local authority will allocate funds for roads, education, water, building and maintenance of council housing, recreation, sewerage, street lighting etc. Public services like the electricity and gas companies will work alongside the council in housing areas allocated by the council planners and built by private companies. There is no unplanned development and provision of public services is uniform throughout the specific area.	Service provision is not uniform across the city. There are some areas which have access to all amenities such as water, electricity, gas, sewerage, waste collection, educational and health services. In other areas people have built their own houses with no access to any of these public utilities.
	Electricity Although nearly 85% of Karachi's population is served by electricity, most districts suffer from regular cuts in supply.
	Sewerage In 1990 only 36.5% of households had direct access to the sewerage system.
	Water In 1990 only 70% of residents received water via direct piped connections.

Table 3.8 A comparison of planning, budgeting and services in UK and Pakistan.

Figure 3.39   Land to be developed in Karachi.

Karachi: squatter settlements

One of the characteristic features of many large cities in developing countries is the existence of large areas of poor quality shanty housing or squatter settlements, often on the outskirts of the city, which lack any of the basic amenities such as piped water supplies, sanitation systems or electricity. In Karachi, these are known as **katchi-abadis** (Figure 3.41). There are 536 listed katchi-abadis with an estimated population of 3 560 800 people, and rising.

They have come into existence partly as a result of the lack of proper urban management and planning, and partly in response to the lack of adequate affordable housing for the poor in the city. The city authority does not have the power to prevent the establishment of such areas.

A typical katchi-abadi

Lyari is an established squatter settlement, situated in the delta of the Lyari River. It has 600 000 inhabitants and is one of the largest settlements.

Figure 3.40   Tanker supplying water to squatter settlements.

Figure 3.41   Shanty houses in Karachi.

The area is becoming more and more densely packed as newcomers build their homes (Figure 3.42). There is no planning at all in the area. It is ideal for jobs as it is so close to the city centre and the Port of Karachi. Because of its ideal location it is sought after by industrialists and speculators. They demolish the shanty houses and build four or five storey tenements with a workshop or shop on the ground floor. Attempts have been made to clear the area by demolition but these have been met with local resistance.

Most of the houses are made from cement blocks but the quality is very poor and they are badly affected during rain or floods. The labourers who live in the area have to walk to work and services are limited.

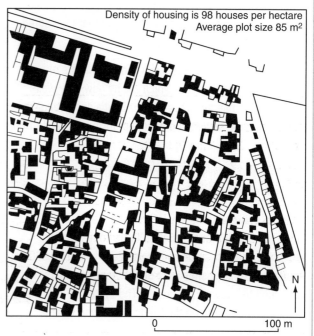

Density of housing is 98 houses per hectare
Average plot size 85 m²

0   100 m

Figure 3.42   Lyari, a city centre katchi-abadi.

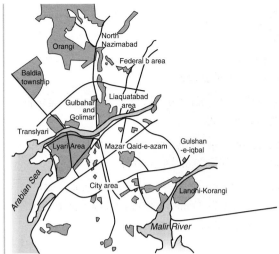

Katchi-Abadis at a glance (Karachi Metropolitan area)			
Area	Number of Katchi-abadis	Acreage	Population
Lyari Area	33	1618	607 000
Translyari	10	550	110 000
Gulbahar and Golimar	37	720	190 000
Baldia township	29	1000	150 000
Liaquatabad area	11	400	60 000
North Nazimabad	12	440	44 000
Federal b area	15	460	50 000
Gulsham-e-iqbal	8	550	60 000
City area	49	1350	160 000
Orangi	28	1800	130 000
Landhi-Korangi	90	1002	87 000
Total	312	9890	1648 000

Figure 3.43   The distribution of katchi-abadis in Karachi.

After a lot of pressure the municipal authorities have put in some water standpoints, some garbage dumps and public latrines. Construction and maintenance of these is very poor, and because they cannot cope with the huge numbers involved they are extremely unhealthy. Around the public standpoints there are pools of stagnant water and mud. Garbage is not collected regularly and lies around in piles. One of the latrines had to be demolished because the smell was more than the local residents could stand.

How do katchi-abadis develop?

Figure 3.43 shows that the katchi-abadis are distributed widely throughout the city area, although there are some concentrations in particular areas.

The Lyari area near Karachi's city centre consists of many older squatter settlements which are by now very well-established, while the areas around Baldia and Orangi Town are more recent developments in the north west part of the city.

There are three ways in which these unauthorised settlements have developed:

1 Illegal sub-divisions
When the city authorities started to try to restrict the growth of squatter settlements, particularly in the 1960s in Karachi, alternative sites were sought by the settlers who continued to stream into the city. The system of illegal sub-division of state land on the fringe of the city became commonplace.

This system revolves around a middleman (or *dallal*), who acquires the occupation of state land by doing deals with local businesses, and then sub-lets plots of land to settlers at prices they are able to pay. He also arranges water supplies and protects the residents from eviction until the settlement is so big that it is safe from that threat.

2 Unauthorised invasions
Many of the earlier katchi-abadi settlements of Karachi arose as migrants to the cities arrived in groups and occupied plots of vacant land as near as possible to the city centre or their work-place. These settlements are therefore very crowded, unplanned with narrow winding lanes, and are often the first targets of attempts to upgrade or improve squatter settlements. Because of the way in which settlers arrived, people tend to stick together in their clan structure, so there is clear ethnic as well as social grouping in these areas. The houses, which are often very similar to the rural village types, have been improved over the years, and services have been built up in a 'do-it-yourself' fashion with the help of the *dallal* and local business people.

3 Organised invasions
In some areas, due to the increase in land prices, people cannot afford to use the *dallal* system to acquire a house plot. As a result, some groups have resorted to what have become known as organised invasions. They earmark a suitable site, occupy it in the evening and build houses on it during the night, then bribe the authorities to prevent demolition and apply for the legal rights to stay.

Only a few such invasions have occurred in Karachi so far, but it seems likely that more will occur in future due to the lack of available land in convenient locations at prices people can afford.

In the early stages of growth the shanty dwellers stand a good chance of being evicted. It is often this threat to their security which encourages inhabitants to join together to resist eviction and to demand the improvement of services.

In the longer-established squatter settlements, especially those where community action has improved conditions, most of the houses are built from bricks or blocks, with proper roofs, and have a reasonable level of facilities such as water, sewerage, garbage collection, storm drainage and electricity. Shanty dwellers are more likely to be given legal rights to their homes if they have been able to provide some basic services for themselves.

The development of a typical squatter settlement in Karachi

Stage 1 (initial invasion)
A few huts appear and are often bulldozed several times before landlords or authorities are forced to recognise the existence of the new settlement. There is no piped water, no sanitation, twisting narrow streets, and no planning.

Stage 2
Families get organised, demand services, and start to improve their dwellings. Water is delivered and sold by private contractors in large tankers. The squatter dwellers will connect their homes to an open pipe system to take sewage away from the area. As services improve there is an increase in the density of population.

Stage 3
More improvements to dwellings take place, including the use of concrete blocks for walls, and the first **standpipes** are located in the area. The houses are provided with electricity from diesel generators.

Stage 4
Drainage and sewerage systems are now installed, usually through self-help schemes, and many houses have water supplies and electricity (at a price). The first shops and two-storey houses are appearing, and the plots are often sub-divided as rents continue to rise.

Questions

4 Why is it so difficult for city authorities to stop the growth of katchi-abadis?

5 Describe the role of the *dallal* in the creation of katchi-abadis in Karachi.

6 Describe the major differences between the planning systems in a UK city and in Karachi.

7 In what ways has the growth of population in Karachi been increased by political factors outside the control of the city authorities?

House types in a katchi-abadi

In one analysis of the squatter housing in Karachi, seven house types were distinguished. These can be seen in Figure 3.44.

Most residents of squatter settlements finance the building of a house from what savings they have, although some are forced to borrow money from friends or moneylenders. In some cases the *thallas* (makers of bricks and blocks), on whom the community depends for building materials, will supply them on credit. This is widespread in the katchi-abadis in Karachi.

Most houses begin with one room and a toilet and this is added to over the years so that after ten years the house may have four or five rooms and it may be more than one storey high. Although masons and other tradesmen are employed by most residents, standards are usually poor and the buildings suffer from defects in structure and poor ventilation.

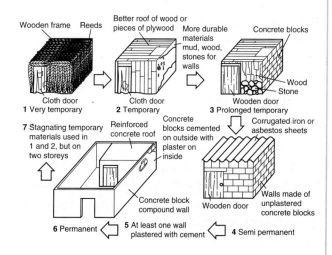

Figure 3.44   House types in a katchi-abadi.

Sewerage is the biggest problem in katchi-abadis, and is often the most neglected service. In Karachi the systems drain into the *nullas* (natural gullies), where the effluent gathers and slowly soaks away. The health hazards from these practices are serious, particularly where the groundwater is contaminated by these soaks.

Health care in katchi-abadis is almost exclusively provided in private clinics, which are usually run by unqualified staff who often use out-dated drugs to treat patients.

Most children are educated in privately run schools, run on a purely commercial basis, and geared to what the residents can afford. They are therefore poorly built, inadequately ventilated and lit, with poor furniture and resources, and staff are almost all untrained, many of them semi-voluntary.

Services in a katchi-abadi

Water

In the early stages of development, water is supplied by *bowzers* (large tankers), usually from the Karachi Metropolitan Corporation via *dallals*. Payments are arranged through the *dallal*, who pockets a profit from the deal. As the settlement grows, these functions are increasingly carried out by local entrepreneurs, who use donkey-carts to deliver water from standpipes directly to customers.

Electricity

Electricity is provided to those residents who can afford it, usually from diesel generators which are installed and operated by local entrepreneurs. The service is only offered for a few homes each evening and on hot afternoons.

Transport

People also begin to make their own arrangements for transport; this includes using small Suzukis to carry passengers to and from work, and to and from local bus stops. The operators have no licence or permits, but have come to an 'understanding' with the local authority representatives (Figure 3.45).

Sewerage

In many areas, it is through the efforts of the people themselves that improvements are made, especially when it comes to the installation of sewage systems. Initially simple tools are used to run small pipelines to the nearest *nulla* (natural gulley) or to a riverbed. However these efforts are often wasted since the pipes are not strong enough to stand up to the use made of them, they are not part of an overall planned system and they quickly block up or fall into disrepair. Since there is an obvious health problem attached to this situation, it is a key area for organised action to take place. The following example illustrates one way out of these difficulties for katchi-abadi dwellers.

The Orangi Pilot Project

In the north west part of Karachi there is a large unauthorised settlement called Orangi, or Orangitown (Figure 3.38). It contains 750 000 people, and has few basic services. In 1980 a self-help programme called the Orangi Pilot Project (OPP) began, with official support, to try to help the communities in this settlement area to develop and improve basic services, particularly sanitation.

The Low Cost Sanitation Programme of the OPP is its most successful effort. Through it the OPP has motivated the residents to manage, finance, operate and maintain an underground sewerage system. The OPP designed the system, worked out its costs, and provided tools and supervision while the residents organised themselves, collected the required funds and managed the implementation of the project. Technical research along with the elimination of contractors, have lowered costs to a quarter of the KMC rates for similar work. As the people have funded the work they also maintain it.

Most of this maintenance is ad-hoc in nature and is in response to a crisis, such as blockage of a drain resulting in flooding of a neighbourhood. In some cases people have, however, developed proper neighbourhood organisations that take care of maintenance and charge the residents a regular fee for it.

Figure 3.45 Transport in Karachi.

However the system eventually flows into the open *nullas*, the development of which is beyond the financial and organisational capacity of local people. The OPP has so far assisted the people in providing underground sanitation for about 60 000 housing units.

Further extensions of all of these services will depend very much on the efforts of the *dallals*, who can often succeed in getting political support (in exchange for promised votes, for example) for improved water supplies and proper electrical connections.

Employment in the katchi-abadis

The residents of squatter settlements have chosen to live there in order to be near to the only available types of employment open to them, which is why most have come to the city in the first place. In Karachi many of the residents of squatter settlements are employed in the chemical and textile industries, and they are used as a cheap source of labour. Many have little knowledge of their legal rights in terms of conditions and wages, and many work in dangerous surroundings. Many of the women gain work in the garment industry, both in cramped factories and at home (doing piece-work). Similarly, shoe-making is carried on in many katchi-abadis, but it is non-mechanised, with little opportunity for profitable expansion.

Families may keep buffaloes and cows, selling the milk produced from them. Some of the women also make fibre mats, ropes and brooms, with raw materials supplied by a middleman, who takes most of the profits. There is a thriving business in the recycling of garbage, with glass, plastics, metals and other materials being collected mostly by children, who pass them on to more middlemen for distribution to manufacturers.

Questions

8 In what ways is service provision in the katchi-abadis never likely to keep up with population growth?

9 Describe the different ways a katchi-abadi dweller might improve on an original reed hut.

10 Why are there serious health problems in a katchi-abadi?

11 Describe the ways in which services are improved in a katchi-abadi.

12 Describe what the residents have to do to get any services in their katchi-abadi.

13 What are the advantages and disadvantages of the self-help schemes in a katchi-abadi?

14 Describe how the assisted self-help scheme has improved services in Orangi.

15 Describe how the Orangi scheme operates and how it offers a cheaper and more effective method than that normally offered by the municipal authorities.

Urban differences

So far, this case study has concentrated on the problems of the katchi-abadis of Karachi, in particular the Mohammad Nagar area of Orangi township in the north west part of the city. However, Karachi also contains a number of areas of middle class and high-income accommodation, many of these dating back to the period of colonial or military domination in the days before Partition (Figure 3.46).

Figure 3.47 shows the street plan of part of one such area. The Defence Officers Co-operative Housing Society was set up in 1952 as an independent organisation with the aim of re-housing military personnel during the chaotic years immediately after the Partition of India and Pakistan. The Society has the legal right to completely control the organisation and planning of the urban landscape in the area under its jurisdiction. Table 3.8 highlights the contrast between this area and katchi-abadis such as Mohammad Nagar.

There could hardly be a more marked contrast within the boundaries of one city. There are obvious differences in the types and layout of the houses.

Figure 3.46   Upper class housing in Karachi.

Density of housing is 10 houses per hectare. Average plot size 700 m²

0   100 m

Figure 3.47   Defence Society housing in Karachi.

There are also differences in the use of the streets. Newer katchi-abadis tend to have a wider planned street layout than the older katchi-abadis nearer the centre of town.

	Defence Society	**Mohammad Nagar**
Total area (ha)	6.5	4.4
Number of plots	67	230
Number of houses	67	300+
Density (house/ha)	10	62
Public area (ha)	2	1.3
Private area (ha)	4.5	3.1
Average plot area (m²)	700	135
Average built-up area (m²)	285	115
Average garden area (m²)	415	91

SOURCE: MASTERPLAN 2000 – KARACHI DEVELOPMENT AUTHORITY

Table 3.9   Urban areas data.

In Mohammad Nagar however, the streets themselves are little used; this is because no-one owns a car, whereas the norm in the Defence Housing Society is for two cars per family, and a busy four-lane road connects this area to the city centre.

Water in Mohammad Nagar is provided via 13 standpipes, which the inhabitants had to pay for and install themselves, after waiting five years for permission to do so from the authorities. In the Defence Housing Society water is piped into every house, and daily water consumption there amounts to 80 million litres per day.

There are no medical or educational facilities at all in the Mohammad Nagar area, and very few throughout the whole of Orangi Township. The cost of travel to other areas is prohibitive, so few children go to school, and most people are forced to receive medical help from illegal practitioners.

Improving housing in Karachi

Any improvement in the housing situation in Karachi is likely to be achieved by co-operation between the municipal government and the residents themselves. The Pakistan government in the 1960s and 1970s did not have a clear housing policy and did not plan housing programmes with a set amount of finance. During this time there were some ad hoc projects. Slums were cleared and low cost flats were built. But even these were beyond the reach of the poorer sections of society, and many were built too far away from available jobs. Such projects were not well planned as far as finance was concerned. The low-income groups could only afford them if they were subsidised and this in turn used up a large amount of the budget and did not provide enough return to finance future projects. There was no way this sort of planning could continue in the long term, as all it did was further drain a limited budget. Some of the projects are listed below:

1950–1951 Karachi Improvement Trust was formed to deal with the different urban problems.

1958 A martial law form of government was established which decided to move the poor out of the city. **The Greater Karachi Resettlement Plan** aimed to resettle 300 000 refugees in high rise flats. Only 10 000 were resettled by 1964 after which time the plan was shelved.

1968 A master planning exercise was initiated and the study revealed that 1.5 million people belonged to low income groups.

1974–1985 The Karachi Master Plan recommended the provision of 91 000 housing plots by 1980 through 'Metroville Housing Project'.

1978–1981 The development authority developed three major townships containing a total of 164 891 plots.

The inadequate and inappropriate response of government policies towards housing problems led to the creation of informal sector housing, with the following problems:

- absence of legal titles to land
- delays in regularisation of ownership
- inadequate physical and social **infrastructure**
- dilapidated housing conditions
- evictions
- danger from disasters, especially in the flood prone areas or low lying lands
- absence of representation at the policy making level.

Karachi 2000 Project

The KDA realised it did not have the expertise to produce a development plan for land use in Karachi. Help was sought from the United Nations Development Programme and the World Bank. The UNDP provided six experts and a joint American-Czech planning firm was contracted to execute the planning work under the title of **_Karachi 2000_**. They drew up three basic housing development programmes which were recommended for the lower income groups in the katchi-abadis.

1. Improvement and Regularisation Programme (IRP)

This was the plan for the lowest income groups, and involved the following:

- a survey of all unauthorised katchi-abadis to see if they could be improved
- secure land tenure. Most of the katchi-abadis were built on land belonging to the municipal authorities. If improvement could be carried out land ownership could be transferred to the residents. This would give them security so that they could further improve their homes without fear of eviction
- improvement of public utilities to a basic standard
- development of a financial plan to achieve the first three aims.

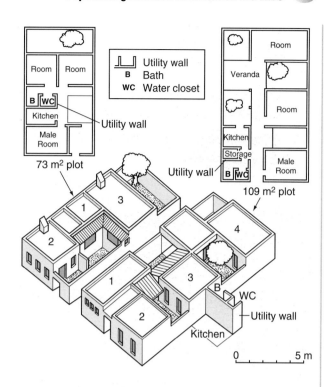

Figure 3.48   The utility wall scheme.

2. Open Plot Development (OPD)

This scheme would provide:

- $72m^2$ plots
- secure land tenure
- public utilities
- community facilities (schools, health centre)
- workshops
- market place.

It provides the initial plans and street layouts and then the new residents build their own houses within the planned area.

3. Utility Wall Development (UWD)

This is a scheme which has been tried with some success in Bolivia and Brazil and is directed towards middle income groups. In each plot of land the municipal authorities would provide a concrete plinth and a concrete core wall along with water supply and proper drainage. It would be up to the new residents to fit the toilet, bath and kitchen facilities and then add on the rooms to this core wall as and when they could afford them. This was a very flexible scheme and offered the way forward for many thousands of families (Figure 3.48).

Unfortunately, this plan was restricted by the lack of finance and only a few of these UWD schemes actually succeeded.

Questions

16 What are the major differences between housing and services in the middle/upper class areas and in a katchi-abadi?

17 For Karachi briefly describe the planning methods which have been used to try to solve the housing problems.

18 Why is transferring land ownership to a katchi-abadi resident seen as a first step to housing improvement?

19 Why is the OPD plan a good idea for katchi-abadi residents on a low income?

Improving transportation in Karachi

Karachi is a major industrial centre and port, and it therefore suffers from a great deal of traffic congestion in the city centre. In order to try to reduce this congestion two possible solutions are being addressed – the building of by-passes and the increased use of public mass transportation.

Two by-passes are planned, one to the north will provide a faster alternative route from the ports to the Super Highway, and one to the south will cut through Korangi (a major industrial area) and join the National Highway.

Karachi's mass transportation facilities mainly consist of buses, railways and minibuses. The bus system is the most important component of the city's transport system (Figure 3.49). Buses at present are being operated by public as well as private operators. The public operators favour the standard single deck buses while the private operators are running mini buses in addition to the standard single deck buses. **The Karachi Transport Corporation (KTC)** has the responsibility for the public bus services while the **Karachi Bus Owners Association (KBOA)** provides the private bus services in Karachi.

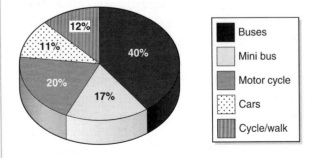

Figure 3.49   Transport modes in Karachi.

Private car ownership increases around 18% every year, as opposed to a 12% growth rate in public buses. There are 37 private cars for every 1000 people compared to 1.4 buses or mini-buses for the same number of people.

The measures taken by the authority to attempt to improve this situation are:

- improve the existing bus system
- construct a new metro rail system in a ring around the CBD
- build an exclusive transit way in selected corridors for public transit modes
- improve rights of way for public transit modes in central locations of the city
- improve the institutional arrangements to support the mass transit programme.

The success of these schemes (which involve large amounts of public funding) will depend on the middle income groups giving up the use of their cars and taking to the buses or trains. This is not likely to happen to any great extent in the near future.

Improving infrastructure in Karachi

Water supply and sanitation

Water is the most essential service and according to the estimates only half the total population of Karachi receives an adequate water supply. Industrial and commercial users are frequently affected by water shortages and water losses during distribution. A large number of household water pipes are undersized, blocked or broken thereby limiting the performance of the sanitation system. The problem of using bucket latrines or direct drain discharge is serious in the squatter settlements of Karachi where the bulk of the raw sewage is spilled into the sea through the Lyari and Malir rivers. The Orangi Pilot Project boasts a successful sanitation project.

Innovative Sewerage in Karachi Squatter Settlements: Orangi Pilot Project

Orangi, a squatter settlement located on the outskirts of Karachi had primitive forms of excreta disposal, poorly laid drains and no rainwater drainage. This created a dangerously unsanitary environment that led to disease, death, and damage to housing stock. Dr Akhtar Ahmed Hameed Khan a world renowned community organiser realised the possibility of installing low income systems for sanitation. The Orangi Pilot Project (OPP) started using the organisational capabilities of the local leadership in each lane.

The modern sanitation system was broken down into its various components:

- sanitary latrine inside the house
- underground sewerage lines with manholes and house connections in the lane
- secondary or collector drains
- main drains and treatment plants.

The emphasis of the Orangi Pilot Project's (OPP) approach was to drastically reduce the cost of construction and to persuade houseowners to accept full responsibility for implementing it. The design of manholes and shuttering was simplified for it to be installed by the owners themselves. The 'line manager' acted as the contractor and represented about 15 houses and the lane became the basic unit of planning. The lane managers were responsible for motivating the households, collecting the money, receiving tools from the OPP and organising the work.

In the beginning the lanes located close to the drain or the creek participated in the programme. The lanes away from the creeks had to come together to lay secondary drains to reach the creeks. This was difficult and took much longer. In the meantime sewage lines clogged up occasionally and had to be cleaned out. To overcome this problem it was decided to put a one chamber septic tank between every connection. This mini septic tank prevented solid matter from flowing directly into the drains. The size and design of the septic tank were determined by its cost to the user and not by any engineering standard. Secondary drains were eventually introduced once more of the community participated in the programme and it became feasible to link all the lanes into it.

SOURCE: AKHTAR BADSHA, SUSTAINABLE AND
EQUITABLE URBAN ENVIRONMENTS IN ASIA

Solid waste

About 6000 tonnes of solid waste are generated in Karachi Metropolitan Corporation (KMC) every day. Only 30% of the volume is collected and disposed of each day while the remainder is either burned or finds its way onto vacant plots, drains or roads.

Karachi is divided into ten solid waste collection districts of which seven are served by KMC while the remaining are taken care of by the airport, defence housing authority and the Karachi Development Authority. KMC spends about 30% of the municipal budgetary allocation for solid waste removal.

The Karachi development plan 2000 stresses the involvement of the private sector in collection and disposal services, and aims to promote recycling and the re-use of landfill sites. A project to dispose of garbage through a garbage train is also underway.

Questions

20 How would traffic congestion in the CBD of Karachi be reduced by the construction of by-passes?

21 Why do you think middle-class people are unwilling to stop using their cars to commute to work?

22 Describe the problems facing Karachi in terms of providing an adequate supply of water and sanitation facilities.

23 Describe some of the advantages of the sewerage system adopted in the Orangi Pilot Project.

Key terms and concepts

Summary

Having worked through this chapter, you should now know:

- the distribution of urban concentrations and the patterns of urban growth and development in Canada and Pakistan
- the physical and human factors involved in the location, development and structure of Toronto and Karachi and their respective land use patterns
- the nature of urban change in these cities, particularly:
  - the causes and consequences of population growth, decline and redistribution
  - industrial, service and employment change
  - housing change, particularly in inner cities and shanty development
  - changes in the quality of city life and the environment
  - the inequalities which may arise in large urban concentrations
- the importance of managing urban change in Toronto and Karachi and the problems which can emerge as a result of economic, social, technological and political processes
- the need for planning controls such as greenbelts, traffic management schemes, and improving the environment of shanty towns
- environmental conflicts within and around the urban areas of Toronto and Karachi due to urban expansion, traffic problems, housing problems and industrial location.

Development and Health

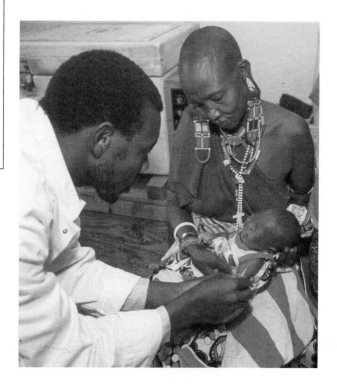

Links to the core

This section involves identifying and explaining differences in social and economic development both between and within countries. It also looks at levels of health, health care and the incidence of disease in the developing world. It focuses particularly on the theme of population and builds on the ideas and concepts introduced as part of the of the human core such as population geography and rural geography. Since environmental issues such as climate, are often significant in explaining differences in levels of development, aspects of the physical core are integrated into the case studies.

What do we mean by development?

A general definition is that development is a process of change over time. The traditional view was that development was synonymous with economic growth i.e. countries and societies advanced by becoming richer. This view was the product of Western attitudes to growth. The emphasis for development was, therefore, on rapid economic growth in poorer nations so that the development, or economic gap, between them and richer nations could be closed.

The modern view states that development should not be measured in solely economic terms. Account should be taken of improvements to people's lives in other ways i.e. in social terms by improving standards of living, diet, access to education and health care; in political areas by moving towards greater social justice, freedom of speech etc. All of these allow people to achieve their full potential by giving them greater choices in their lives. Allied to this is the idea of **sustainable growth**, i.e. that resources are used responsibly, not exploited until exhausted so that the needs of future generations become part of the development equation. In this view development is a complex process involving economic, social, political, environmental and cultural change.

Indicators of development

In the traditional view indicators of development were based on economic indicators only e.g. **Gross National Product (GNP)** per capita (the total value of economic output divided by the population). This includes figures for food production, value of goods produced, provision of services, profits from overseas investments and money earned from foreign business. Using such an indicator, Figure 4.1 shows the distribution of high, middle and low income countries in the world.

High income $8500+

Middle income $700-$8499

Low income $699 or less

Figure 4.1   Global distribution of GNP per capita.

Figure 4.2   A rural landscape in the Developing World.

As would be expected, high income countries are found in North America, Western Europe, Japan, Australia and New Zealand whilst the poorest countries are concentrated in Africa and south east Asia.

The major advantage of using an indicator of this type is that it is relatively easy to calculate. The underlying assumption here is that if a country is wealthy the quality of life for all of citizens will automatically be good.

There are, however, a number of disadvantages associated with the use of economic indicators. As the figures involved are averages, they disguise inequalities between the richest and poorest in society, and also between different regions within the country. The figures exaggerate the **Development Gap** (the difference in development between countries) as they do not take into account **relative purchasing power**. As the costs of goods and services tend to vary from country to country, a set amount of money may buy more goods and services in one country as compared to another where costs are higher. Economic indicators also ignore **subsistence activities** such as

farming and the informal economy, despite the fact that both of these are the basis of economic activity in rural economies in the developing world (Figure 4.2).

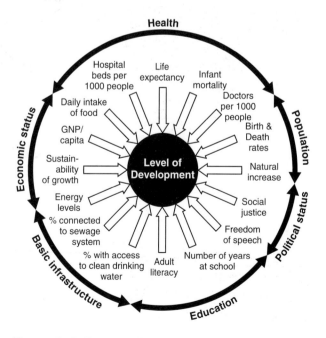

Figure 4.3   Indicators of development.

Using economic indicators to compare different countries also presupposes that currency values remain relatively stable in relation to one another e.g. £100 will always be exchanged for $50 or 60 Euros. But this is not the case. Currency values change on a daily basis, usually only by a small amount, but occasionally by significant amounts, so that comparisons based on monetary values alone are often inaccurate. In addition, conditions peculiar to a country such as long cold winters or large distances to be travelled are not taken into account. These can influence inputs to the GNP in the form of high fuel costs or more expensive goods and services. Finally economic indicators imply that the only way to improve people's lives is to go for Western style economic growth, ignoring other factors which might influence a country's growth.

The modern view believes that there is a need to take into account a much wider range of data, some of which is extremely difficult to quantify, in order to give a more rounded view of the quality of life within a country. There are a great variety of such indicators as shown in Figure 4.3. These include statistics on health such as life expectancy at birth, infant mortality, numbers of doctors and hospital beds per thousand of the population and the daily intake of food. Education levels are shown by literacy levels and number of years in full time education. Both of these groups are valuable in showing the ability of a country's government to supply basic services to its people. Population characteristics, birth and death rates and natural increase, indicate the rate at which the population is growing. A high growth rate could cause severe problems for a country struggling to develop. Economic status is shown by indicators of industrial and agricultural output, the percentage employed in the primary, secondary, tertiary and quaternary sectors and the rate of energy use. The ability of the government to supply basic infrastructure services is shown by access to clean drinking water and the number of people connected to sewage systems. The levels of political freedom can be shown by statistics on freedom of speech and access to justice within the country.

Statistics on education standards and health care provision are relatively easy to obtain. It is much more difficult to measure such factors as freedom of speech, or to assess whether growth is sustainable.

Physical Quality of Life Index

- 90 or above
- 78–89
- 56–77
- 31–55
- 30 or below

Figure 4.4   Global distribution of the Physical Quality of Life Index.

Human Development Index
- 0.90+
- 0.75-0.899
- 0.500-0.749
- 0.25-0.499
- 0.048-0.249

Figure 4.5   Global distribution of the Human Development Index.

Three indicators have been regarded as especially significant:

- **life expectancy at birth** – this provides a good reflection of the quality of health care and levels of hygiene and sanitation available, and which reacts quickly to improvements in the provision of such services
- **access to safe water** – this indicates the quality of the provision of basic services within an area and is especially useful when studying urban/rural contrasts
- **adult literacy** – this is a good indicator of the levels of skill and expertise within a country which would allow it to move forward in its development.

What is important, however, is that a *single* indicator should never be used to determine development as a country might score well in one but badly in another. It is always best, therefore, to use composite indicators.

Composite indicators of development

There are now a number of **composite indicators of development**.

One of the first to be used (in 1977) was **PQLI – Physical Quality of Life Index**. It combines figures for **life expectancy**, **literacy rates** and **infant mortality rates** so that it concentrates on social rather than economic development and was, therefore, useful as a counterbalance to the emphasis on economic development prevalent at the time. To calculate it, figures for each of the three factors for a country are entered on a scale from 0 (worst figure) to 100 (best figure). The three figures are then averaged to give the country's PQLI. A figure of 77 and above is considered satisfactory, below that the quality of life is regarded as poor. As can be seen from Figure 4.4 the PQLI distributions broadly correlate with GNP distributions although it does throw up some interesting anomalies. For example several middle income countries such as Argentina and Chile score better than would be expected, while others such as Saudi Arabia score far worse than many of the countries of Africa and south east Asia. These reflect the different emphasis and spending priorities which these countries give to social policies.

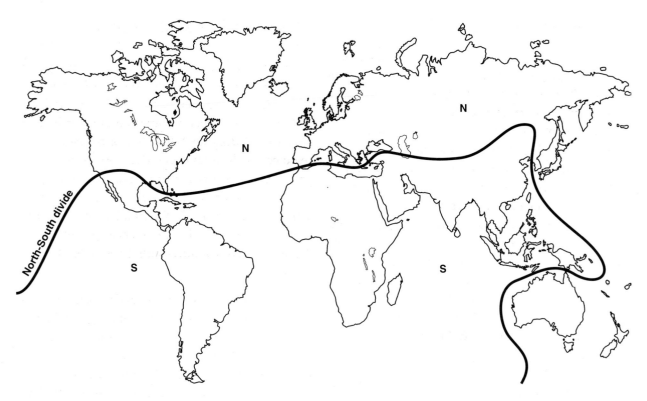

Figure 4.6   The North–South Divide.

A more recent composite indicator is the **Human Development Index (HDI)** which was first used in 1990 by UNPD (United Nations Programme for Development). This attempts to combine a range of economic and social indicators.

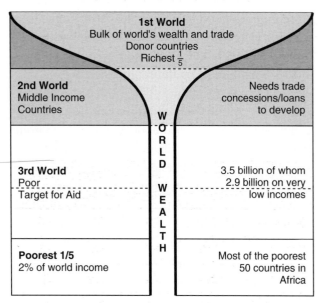

1st World	
Bulk of world's wealth and trade Donor countries Richest $\frac{1}{5}$	
2nd World Middle Income Countries	Needs trade concessions/loans to develop
3rd World Poor Target for Aid	3.5 billion of whom 2.9 billion on very low incomes
Poorest 1/5 2% of world income	Most of the poorest 50 countries in Africa

Figure 4.7   Classification of countries according to level of development.

The Human Development Index uses **adjusted income per capita** (i.e. the purchasing power of a given sum within the country), **educational attainment** (a combination of adult literacy rates and average number of years of schooling) and **life expectancy at birth**. The best level is 1 and the poorest is 0. A country's score for each indicator is ranked between these, and the figures added and averaged. A major advantage of this indicator is that it allows direct comparisons to be made between countries. It can also be adjusted to show figures for male and female incomes. This gives interesting results as some otherwise high ranking countries, such as Japan and UK, slip down the rankings reflecting the disparity between average male and female earnings in these countries. Figure 4.5 shows the distribution of HDI values and indicates a broad correlation with GNP but also highlights similar anomalies to that of the PQLI index. There are limitations as only three sets of data are used. The results still mask internal differences and as these are relative figures, if all countries improve at the same rate the poorer countries will never climb up the rankings and will get no credit for their achievements.

Why measure development?

It is only by measuring development that it is possible to see what progress is being made towards improving the quality of people's lives. It helps to focus attention on areas where there is a need for more encouragement, and areas where aid from donor countries could be targeted. Comparisons can be made on a global scale, allowing analysis as to why some areas have progressed while others have not. The use of development indices has led to the classification of countries according to their perceived levels of development, for example the **North–South Divide**, first described in the Brandt Report (Figure 4.6). Another classification used is that of **First World** (the industrialised Western democracies), **Second World** (the centrally planned economies of Communist states), and **Third World** (poor, underdeveloped areas mainly in Asia, Africa and South America) as shown in Figure 4.7. These classifications have had to be modified in the light of changing world circumstances. For example the collapse of the centrally planned economies of the Soviet Union and Eastern Europe, the rise of the wealthy oil states in the Middle East and the emergence of the **NICs (Newly Industrialised Countries)** especially the 'Tiger economies' of south east Asia (Figure 4.8).

Contrasts in development

There are many reasons for contrasts in development. Some countries such as Saudi Arabia or Brunei have been able to rely on income from their considerable oil and gas reserves. Other countries, for example Ethiopia, Somalia and Bangladesh have few or no natural resources and have experienced severe climatic problems such as drought or devastating flooding. In other areas such as Sudan and Sierra Leone, civil war and other forms of political instability affect their ability to improve their level of development. Table 4.1 shows countries with differing levels of development.

There can also be contrasts in development within an individual country. Statistics which are given for a country are average figures. As such they disguise differences in levels of development within a country so that care has to be taken in using such statistics. There may be a few very wealthy families with the majority of the population living at subsistence levels. Regional differences also exist with some areas more well off than others. Rural and urban contrasts are common in developing countries. Remote, isolated and scattered rural populations are less likely to have access to services such as clean water and sanitation than those in crowded urban areas. Also rural populations are less likely to have access to education and health services. In times of crisis rural populations are less likely to receive help from international aid agencies than those in the more accessible urban populations.

Q u e s t i o n s

1 Explain the differences between the traditional and modern views of development. How does this affect the choice of indicators used in each case? What are the advantages and disadvantages of the indicators used in each case?

2 For the PQLI and the HDI or any other composite measures of development which you have studied, state three indicators which might be used in their calculation, commenting on their usefulness.

3 Explain why it is important to measure levels of development despite the problems associated with the collection of such data.

Country	GNP per capita $	Life expectancy	% Adult literate	% Infant mortality
Singapore	26 730	76	91	4.0
Vietnam	240	67	90	38
Saudi Arabia	7040	70	16	29
Bangladesh	240	58	33	77

Table 4.1 Contrasts in development.

NICs

An NIC is a Newly Industrialised Country (Figure 4.8). The term was coined in recognition of the significant changes in the level of industrialisation in a number of developing countries. A NIC is defined as a country in which the level of **Gross Domestic Product** (the value of all goods and services produced in the country) *from industry* exceeds one third of the *total* GDP. These changes are often due to a rapid expansion of investment by Transnational Corporations (large international businesses with headquarters in one country and a manufacturing presence in many others e.g. Shell, IBM). Included in this category are such countries as Brazil and the Asian 'Tiger economies', i.e. Taiwan, Hong Kong, South Korea and Singapore. The latter four have experienced rapid and sustained growth year on year for the last 25 years. A second wave of growth has recently occurred in south east Asia in the economies of Indonesia, Malaysia, Thailand and the Philippines which are attempting to emulate the original 'Tigers'.

Why have the Asian Tigers been so successful?

Some of the Asian Tigers have considerable natural resources e.g. tin and rubber in Malaysia and coal, oil, natural gas and forest products in Indonesia. In the past, these were exploited as primary exports but have more recently provided a base for manufacturing industries as the country developed. Other countries such as Singapore had no natural resources apart from its harbour. However, their success was not primarily based on natural resources.

All of the Tiger countries had certain assets in common. These included a large, flexible, hard working, low cost labour supply. Based on this it was possible to develop low skilled, labour intensive industries such as textiles, clothing, toys, plastic, leather goods and footwear.

There was also strong government intervention to encourage and help develop export orientated manufacturing. Government policy was mainly geared towards attracting foreign investment in order to introduce high tech industries. The policy was very successful in attracting **Transnational Corporations** which appreciated the low labour costs, reduced costs of production and growing local markets, as the countries themselves became more prosperous.

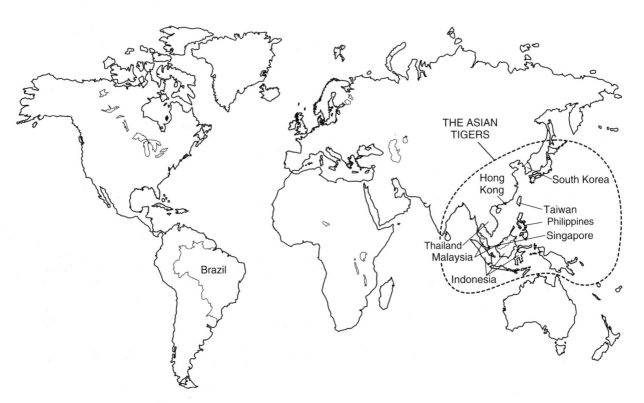

Figure 4.8   The NICs.

They were especially attracted by the proximity of the growing Chinese market, lack of trade union activity and strikes, inducements from the government such as reliable power supplies, an efficient infrastructure and relaxed laws in such areas as pollution control and employment law.

In addition, local people were encouraged by government policies to save a relatively large part of their earnings so that there was considerable capital generated internally between the 1960s and the 1990s. As a result they could become major investors in their own economies to help move manufacturing forward into sophisticated, high value new products, mainly in electronics, telecommunications and aerospace.

Consequently these countries have increased domestic production, decreased poverty, developed a high skill base and a higher living standard than many of their neighbours. The rapid growth did, however, lead to the exploitation of labour, especially of women and children, and the growth of illegal immigration from their less developed neighbours looking for work and a chance to improve their quality of life. These groups usually got the dirty, dangerous jobs with no legal protection which the local people were no longer prepared to do. The rapid expansion also caused considerable environmental damage through deforestation, land degradation and air and water pollution.

A heavy dependence on foreign investment means that these economies are highly vulnerable to shifts in the global economy, and as a consequence they have suffered badly in the world recession of the 1980s and 1990s. In South Korea, Thailand, Indonesia and Malaysia currency values plummeted and manufacturing output was badly hit during this time. The position was made worse as overborrowing on the world money markets led to large debts, and this, combined with the collapse of the internal banking system, led to a melt down in the value of the currency in the international echanges. A further problem is that corruption is common at all levels within these economies and **cronieism** (where contracts are awarded to relatives and friends regardless of their competence) is a widespread practice.

Case Study Singapore

The foundation of the modern state of Singapore dates from 1819 when the island was chosen by Sir Stamford Raffles to become the centre for the British East India Company due to its natural harbour and strategic position on the trade routes between India and China. Trade was further boosted by the opening of the Suez Canal and the development of steam ships (Figure 4.9).

Singapore became the centre for the export of rubber and tin from Malaysia. It grew steadily as a trading centre for the region. After independence in 1959, the new government began a crash programme to develop its manufacturing base which proved to be a remarkable success story. In the 1960s it concentrated mainly on job creation, developing labour intensive industries such as textiles and shipbuilding.

By the 1970s, Singapore had achieved full employment and began to shift the economic emphasis to the creation of a highly skilled work force producing high value goods and to the encouragement of foreign investment.

Industries included shipbuilding and repair, financial services, tourism, high tech industries (electronics, aerospace components and telecommunications equipment), textiles, chemicals, plastics, oil and petrochemicals.

Figure 4.10   Singapore skyline.

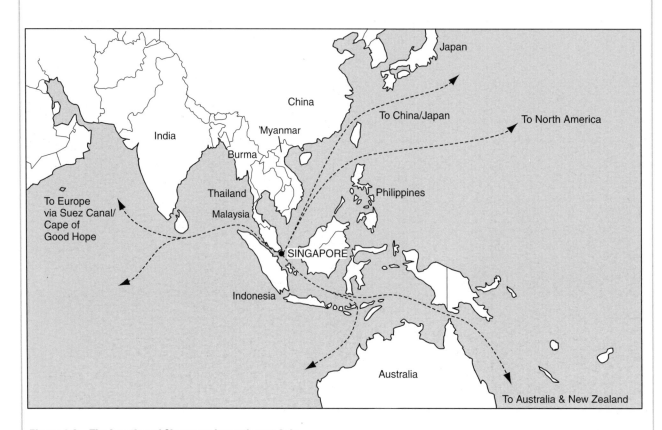

Figure 4.9   The location of Singapore in south east Asia.

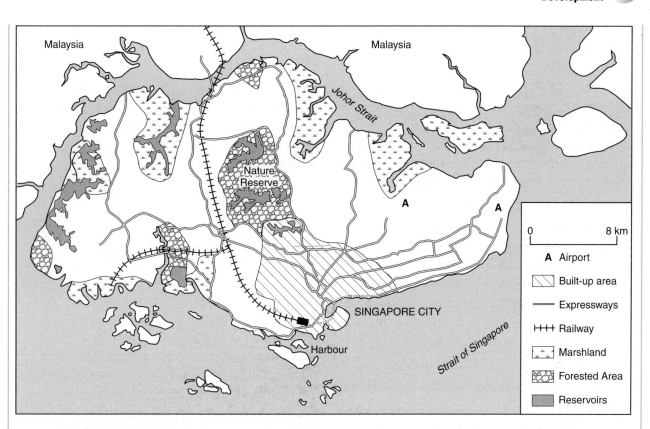

Figure 4.11   Singapore in the 1960s.

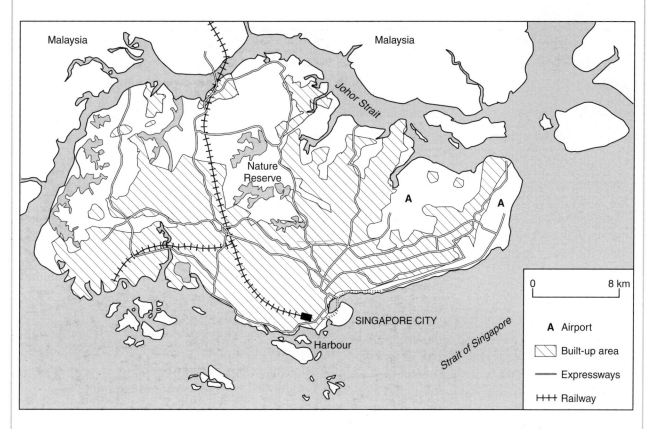

Figure 4.12   Singapore in the 1990s.

The early rapid economic growth of 8% per annum is now slowing as Singapore is facing competition from neighbouring states. Its citizens have a high standard of living and per capita incomes above those of Germany.

All this has been made possible in a tiny country with few natural resources by a well educated, hard working population, low population growth, excellent sea and air links and above all a stable political situation with a highly interventionist government which fosters a strong sense of national identity. The government gives strong direction towards development by creating free, open markets and ensuring that there is little corruption or crime and few strikes. This has created a favourable climate for foreign investment which is welcomed and made as free from regulation as possible (Figure 4.14).

The government's targets for the future are that Singapore becomes the new centre for international trade, and a global centre for financial services, high tech industries and business services ranging from research and development through to manufacturing and distribution.

The country has remained stable, although the government is under pressure to free up its grip on society.

Figure 4.14   The skyline of the central financial district of Singapore.

People remain unsure of the consequences of this. The economy is also vulnerable to the breakdown of government in neighbouring states such as Indonesia, which could disrupt trade. The economy is now having to face the consequences of success in the form of the high cost of land on a crowded island and the increasing costs of a highly skilled labour force. A very strong currency is making exports more expensive, and the economy has become heavily reliant on foreign investment.

The government has tried to counter act some of these problems by encouraging local companies to locate new industrial development in lower cost neighbouring states e.g. Malaysia, Indonesia, China, India and Myanmar, by building industrial estates there. Its broader, more established industrial base and proven ability to manage change has, to date, enabled it to weather the financial crisis in south east Asia better than many of its neighbours.

Figure 4.13   Tankers in Singapore Harbour.

Less economically developed countries

The position of the NICs is in marked contrast to the situation in many countries of Asia and Africa. There, economies have stagnated, grown slowly or even declined.

In parts of Africa growth rates have plummeted, food production has decreased, incomes per head have fallen and overall, industrial performance has declined. It is in fact the only region of the world where per capita output declined consistently throughout the 1980s and the quality of life of its citizens deteriorated. This is highlighted by the fact that the bottom 9 countries out of 175 in terms of HDI are all in Africa.

For both the **World Bank** and the **International Monetary Fund (IMF)** efforts to try to deal with this crisis became the focus of attention. Their solution was to insist on **Structural Adjustment Programmes** which aimed to aid economic progress. The policies tended to switch money from social action to agricultural and industrial development in an attempt to stimulate economic growth. There were also attempts at developing **import substitution** policies where local industries were encouraged to go into production in an attempt to cut down on expensive foreign imports.

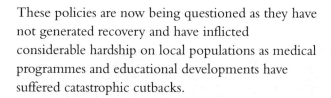

These policies are now being questioned as they have not generated recovery and have inflicted considerable hardship on local populations as medical programmes and educational developments have suffered catastrophic cutbacks.

Causes of the problems in Africa

- Africa is an arid continent where, over the last 30 years, annual rainfall has been well below average. This, in a continent where most people are heavily dependent on agriculture. As a result deforestation, overgrazing, soil erosion and desertification are widespread.

- Africa in general has very high population growth rates, in some countries in excess of 3% per annum. Family planning programmes have had little impact in these areas due to traditional attitudes which value large families. Such population increases tend to wipe out any progress in a country's development.

- Throughout Africa there are high levels of disease e.g. Malaria, intestinal parasites, HIV and AIDS. The scale of the problem has left health systems unable to cope and led to a serious loss of members of the economically active population. There are low rates of education and the small number of Africans who are educated and skilled leave for higher paid employment elsewhere.

- Civil wars have created large numbers of refugees throughout the continent and led to inflated spending on military budgets at the expense of social spending. These conflicts are often a legacy of the area's colonial past when country boundaries were drawn with no consideration for tribal groupings. Such policies created political instability both within and between countries.

- In the past exploitation of Africa's resources by colonial powers for primary exports meant that there was little development of manufacturing industry, and fertile land was used for cash crops rather than food crops. This pattern has proved difficult to change.

- After independence many African countries became involved in misguided, large scale development projects which they had neither the resources nor the expertise to maintain. More recently there has been an unwillingness to be open to outside investment because of memories of past exploitation, despite their inability to generate capital for development by themselves. All these problems have been compounded by unstable governments, poor leadership, high levels of corruption and a widespread disregard for human rights.

- Much of Africa suffers from a poor **infrastructure** with badly maintained road and rail links. There is a lack of natural resources throughout the continent and therefore a need to import commodities such as oil. Economies were particularly badly hit by the dramatic rise in oil prices in the 1980s, a situation made worse by income from their exports falling drastically at the same time.

- Due to heavy international borrowing during the 1960s and 1970s, African countries are weighed down by external debt, and the repayment of these debts has been taking more and more of their foreign earnings, leaving little to enable them to pursue internal development. Recently, the world's most economically powerful nations, the **G8 countries**, have begun the process of cancelling the debt of the worlds poorest countries.

Case Study **Kenya**

In the north of Kenya there are semi arid plains and desert which are prone to drought. In the south, savanna vegetation dominates, with many game parks e.g. Masai Mara. In between are the Central Highlands which are the most fertile area containing 85% of the population so that population densities are very high. The area also includes the Rift Valley a dramatic geological feature 100 km wide and 3 km deep (Figure 4.15).

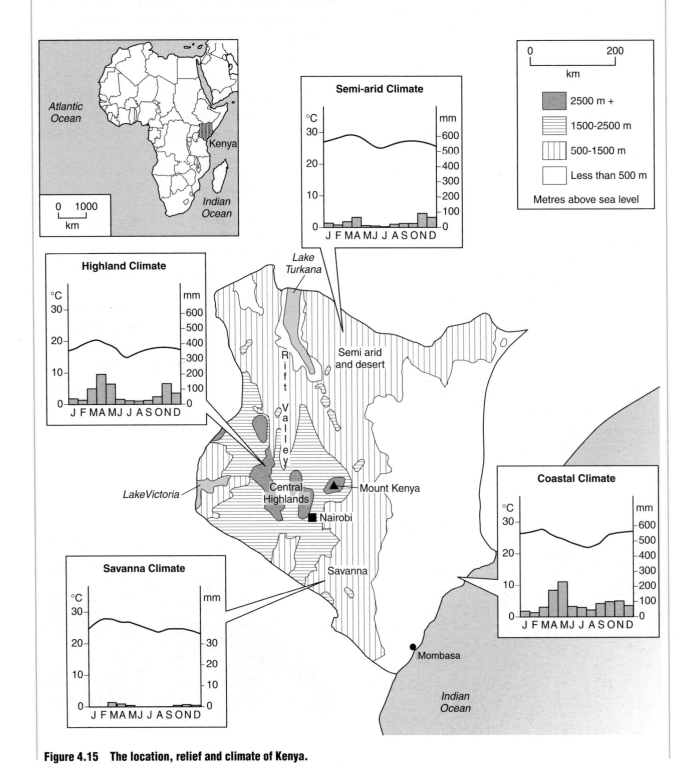

Figure 4.15 The location, relief and climate of Kenya.

Only one fifth of the country is suitable for agriculture, most of which consists of subsistence farms, the main crops being maize and cassava (Figure 4.16). There is limited area available for the expansion of crop land but there is potential to increase yields. In the semi arid areas, livestock farming predominates. A succession of droughts have resulted in overgrazing, land degradation and soil erosion. There is a need to increase the area of irrigated land but this has not been possible because of lack of finance. Often the most fertile areas are used for cash crops such as tea (which provides 28% of Kenya's foreign exchange and one million jobs), coffee (the third most important foreign currency earner), fruit, flowers and vegetables (Figure 4.17). All this puts pressure on food supplies so that Kenya is now a net importer of food. The situation is not helped by Kenya's high birth rates. Tourism is increasingly important to the economy due to the wealth of wildlife and beach holidays on the Indian Ocean coast.

Despite its problems, Kenya is East Africa's most industrialised country although the industry is based mainly in Mombasa and Nairobi. These industries supply both the domestic and export markets, and are based mainly on food processing with some chemical and engineering plants. Growth is hampered by a lack of fuel, Kenya has little coal, oil or gas and relies heavily on imports. Kenya also has to import manufactured goods and has been hit by rising prices for fuels and the falling value of its primary exports. The country suffers from frequent power cuts (as developments in hydroelectric power are affected by droughts). The government is planning to develop geothermal power stations in the Rift Valley area and is also engaged in oil exploration.

Other problems include deteriorating roads, rising rates of crime and violence, corruption, political unrest and high inflation rates. These all badly affect the economy, put off potential investors and cut into the tourist trade.

High population growth rates are making development difficult as they put pressure on education and health services, food supplies and land. Kenya needs to improve its infrastructure; especially roads, airports and water supplies (there are particular pressures on these from tourist developments as well as from the growing urban demand). Disease rates are high especially for HIV and AIDS which reduce the capacity of the workforce in both the agricultural and industrial sectors.

Kenya's economy is vulnerable to adverse weather conditions and fluctuations in world markets, but its government has declared its intention to become an industrial nation by 2020. In order to do so Kenya will have to create conditions attractive to foreign investment. In recent years Kenya's situation has deteriorated markedly. The IMF suspended loans to the country in 1997 because of government corruption, although the suspension was lifted in 2000. In the meantime Kenya suffered from power shortages, inflation, water rationing and a steady fall in the value of the currency. Many farmers were unable to continue due to drought and have moved to the cities. Cash crops have also been hit by mismanagement and corruption.

Figure 4.16   Rural Kenya – a small subsistence farm.

Figure 4.17   Workers in a women's cooperative producing tomatoes for sale.

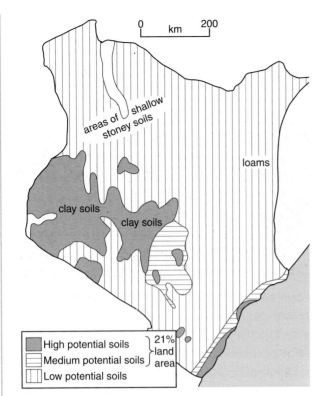

Figure 4.18   Land productivity in Kenya.

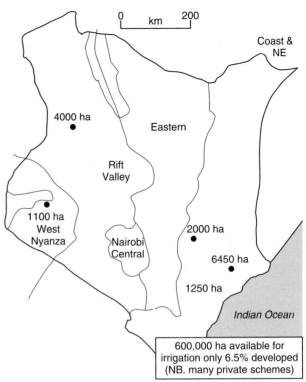

Figure 4.20   Proposed large scale irrigation projects in Kenya.

The tourist industry, once a major earner of foreign exchange, is in decline because of reports of an increasingly violent and dangerous society. Despite increased aid and other international help Kenya's economy is in such decline that it will take considerable time and effort to return it to the situation at the time of independence in 1978.

There are considerable differences in levels of development within Kenya. These differences are due to variations in relief, climate, land use potential (the capacity of the land to sustain agricultural developments) and accessibility (Figures 4.15, 4.18 and 4.19). There are also contrasts between urban and rural areas and within urban areas.

In the north and east of the country, which consists of 80% of Kenya's area, land is mainly below 1500 m and the climate is arid or semi arid with low, erratic, unevenly distributed and unreliable rainfall. There are many areas of shallow, stoney soils (Figure 4.18). All this combines to give a landscape of dry bushland, desert and scrub vegetation with a low agricultural potential where, in the driest areas, harvests fail on average three years out of five.

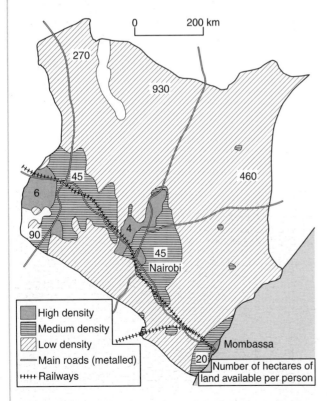

Figure 4.19   Population density and communications in Kenya.

Accessibility in the area is poor with few metalled roads and no railways (Figure 4.19). Population densities are low with approximately six million people living in the area. They are mostly nomadic pastoralists (the Masai) relying on their herds of cattle and goats. Despite the poor quality of the land in the area it makes a valuable contribution to Kenya's agricultural output with 80% of Kenya's cattle found there. Population growth is high at 2.5% per annum so there is considerable pressure on the land from overgrazing and deforestation. The area has suffered badly through a series of droughts in recent years forcing many of the pastoralists to move to the overcrowded urban areas or to subsistence farming (Figure 4.21). The area has long been neglected by successive Kenyan governments. Despite pressure on scarce water supplies, little has been done to increase the area of irrigated land, with relatively few government sponsored schemes (Figure 4.20). There has been little or no attempt to research the agricultural problems of the region. The area is poorly served by health and family planning clinics which are scattered and difficult for many people to access.

In the south and west of Kenya, the land is mainly coastal plain or upland. Rainfall is much more reliable and soils are a mixture of clay and alluvium with small areas of fertile volcanic deposits (Figure 4.18). The vegetation is forest and savanna and in terms of agricultural potential the land is mainly in the medium to high category. Over 28 million people live in the area so population densities are high and there is considerable pressure on land. The area produces the bulk of Kenya's crops for export such as coffee, tea and fruit (Figure 4.22).

Figure 4.21   Masai herders which have been forced to turn to farming.

The area has been the focus of development with large scale irrigation schemes planned (Figure 4.20). In addition, Health and Family Planning clinics are more freely available.

The bulk of Kenyans live in the countryside (Figure 4.19). Only 30% live in urban areas. There are considerable contrasts between conditions in the countryside and urban areas. Almost 70% of urban dwellers have access to clean water (which has to be paid for) compared to 50% in the rural areas. In addition, in most rural areas people have to travel considerable distances in order to collect water. In the urban areas there is access to education (which apart from primary education has to be paid for), doctors and other health facilities.

Figure 4.22   Workers in a tea plantation.

Figure 4.23   Central Nairobi.

Figure 4.24   Urban Kenya – good quality housing.

Figure 4.25   Urban Kenya – slum housing.

Within urban areas there are also contrasts. The centres of Nairobi and Mombasa for example contain tall modern buildings similar to those found in most of the world's cities. There are also areas of good quality housing (Figure 4.24). Surrounding these cities however, are urban slums which are growing rapidly as people migrate to the cities when they are unable to make a living in the countryside (Figure 4.25). In these slum areas water supply is inadequate and many people are unable to afford food or education for their children. Sewage systems are almost non existent, housing is of very poor quality. Disease spreads rapidly in these conditions. Many people have moved into the cities in the hope of finding employment. Unemployment rates, however, are over 50% and people find it difficult to improve their lives.

Questions

4 For two named indicators of development suggest how they might be used to illustrate a country's level of development.

5 How have the NICs been able to develop so rapidly? What have been the disadvantages of this rapid growth?

6 Explain the failure of the African countries to improve their levels of development.

7 Why do indicators of development not always produce an accurate picture of life within a country?

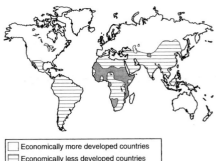

☐ Economically more developed countries
☐ Economically less developed countries
■ 42 least economically developed countries

8 Study the information on Brazil in Figure 4.26. Using the information on Brazil, or any other country you have studied, explain why levels of development vary throughout the country.

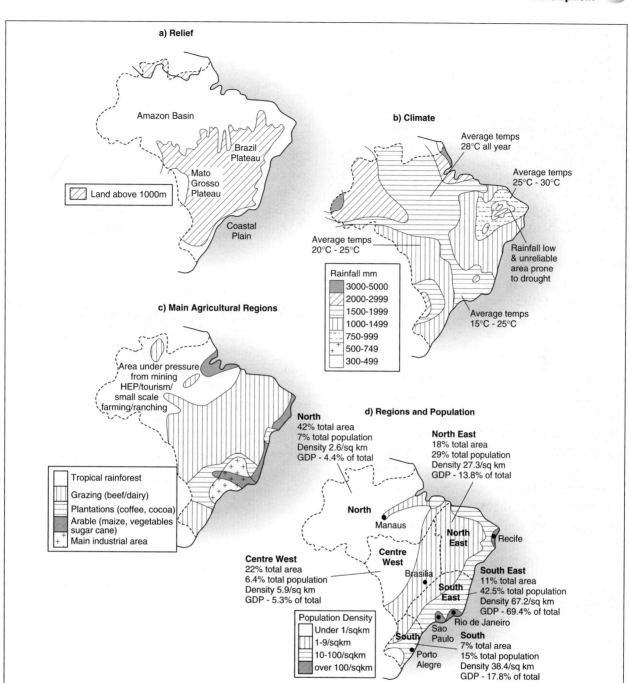

a) Relief

Amazon Basin

Brazil Plateau

Mato Grosso Plateau

Coastal Plain

▨ Land above 1000m

b) Climate

Average temps 28°C all year

Average temps 25°C - 30°C

Average temps 20°C - 25°C

Rainfall low & unreliable area prone to drought

Average temps 15°C - 25°C

Rainfall mm
▦	3000-5000
▨	2000-2999
▤	1500-1999
▭	1000-1499
┄	750-999
+	500-749
	300-499

c) Main Agricultural Regions

Area under pressure from mining HEP/tourism/ small scale farming/ranching

☐	Tropical rainforest
▦	Grazing (beef/dairy)
▥	Plantations (coffee, cocoa)
▨	Arable (maize, vegetables sugar cane)
+	Main industrial area

d) Regions and Population

North
42% total area
7% total population
Density 2.6/sq km
GDP - 4.4% of total

North East
18% total area
29% total population
Density 27.3/sq km
GDP - 13.8% of total

North
Manaus

North East
Recife

Centre West
22% total area
6.4% total population
Density 5.9/sq km
GDP - 5.3% of total

Centre West
Brasilia

South East
11% total area
42.5% total population
Density 67.2/sq km
GDP - 69.4% of total

South East
Rio de Janeiro
Sao Paulo

South
Porto Alegre

South
7% total area
15% total population
Density 38.4/sq km
GDP - 17.8% of total

Population Density
☐	Under 1/sqkm
▥	1-9/sqkm
▤	10-100/sqkm
▦	over 100/sqkm

Figure 4.26   Brazil: relief, climate, agriculture and population.

Factors involved in the incidence of disease

The incidence of disease is due to a combination of physical and human factors. These vary in importance throughout the world and account not only for the prevalence of disease but also for the different types of disease dominant in an area.

Physical factors

Climatic conditions have a considerable impact on the level of health. Hot, wet conditions encourage the spread of infectious diseases and provide ideal breeding conditions for insects which act as vectors (intermediate hosts which transmit disease but are not affected by it). Drought and arid conditions affect water supplies which in turn can result in increased levels of disease as water supplies become restricted and/or contaminated.

The quality of the water supply is frequently a determining factor in the level of health within a community. In the developing world, on average 50% of households lack access to clean water supplies, although this percentage rises dramatically in rural areas. In urban areas access to piped water supplies is easier to provide. In rural areas people may have access to a well deep enough to supply clean water all year round. More often, however, the well will be too shallow, liable to dry up in the dry season or become contaminated by sewage. For the poorest people, the water supply may be a considerable distance away and consist of a stagnant polluted river or pool. People, mainly women and children, have to make several daily trips on foot to carry this water back to their villages. This is time consuming and can have implications for the health of those who have to carry the water each day.

Related to the poor provision of clean water supplies is the lack of **basic sanitation** in the developing world.

Few people, even in urban areas, have access to sewage systems. Because of this, diseases caused by contact with sewage are widespread in both urban and rural areas. Children are particularly at risk from diseases spread by contaminated water supplies. For example, four million children under five die of diarrhoea each year caused by having to drink contaminated water, despite the fact that the illness is preventable.

Mountainous areas usually have poor communications and are remote and isolated. In such communities, access to health care and medical services, even of the most basic kind, is difficult. As a consequence people's health levels are generally poor.

Human factors

A major cause of poor health and low levels of health care is poverty. Poverty means that many people live in overcrowded slum conditions lacking running water and sanitation. Others cannot afford housing and live on the streets. In such overcrowded, insanitary conditions disease spreads rapidly. Poverty leads to poor levels of education so that people are ignorant of the causes and treatments for even the most preventable diseases. Developing world countries are often unable to afford the mass immunisation programmes which have been so effective in reducing disease and death in the developed world.

Poverty also leads to poor levels of nutrition as people are unable to afford a healthy diet. Even in areas where people support themselves by farming, poor farming methods, overuse of the land and consequent soil erosion lead to yields which are too low to support the population. People who are undernourished are more susceptible to disease and once ill find it harder to fight their illness. As a result of all this, over one billion people (one fifth of the total world population) lack adequate food, clean water, elementary education and basic health care.

World diseases

The lack of available clean water and adequate sewage systems are major factors in the incidence of many of the most common diseases in the developing world. These can be classified as:

- **water related** – diseases transmitted via an insect vector breeding in the water e.g. malaria and river blindness
- **water based** – diseases transmitted via parasites using hosts living in the water. e.g. schistosomiasis
- **water borne** – diseases transmitted by people using water contaminated by sewage e.g. cholera and typhoid.

Malaria

Malaria affects 200–500 million people worldwide each year. In the past 15 years the disease has killed nearly 50 million people, with 2 million deaths in Sub Saharan Africa in 1999 alone. 85% of the deaths occur in children with one child dying every 20 seconds. Those who survive develop a limited measure of **immunity** in adulthood. They can however be reinfected, suffer from recurrent and violent bouts of fever, and are severely anaemic.

This weakens their immune system so that they are less resistant to other diseases. The disease also has adverse economic effects in areas where it is endemic (part of the background to life). This is particularly damaging in areas where agriculture is a significant component of the GNP. It has been estimated that a single bout of fever causes the loss of 10 working days.

We tend to think of malaria as a tropical disease associated with developing countries but this is not so. Apart from hot and cold deserts and high altitude areas mosquitos have a worldwide distribution (Figure 4.27). In general the incidence of the disease is found between 65°N and 40°S and below 3000 m. With global warming and rising average temperatures disease carrying mosquitos may spread northwards into America and southern Europe.

The malarial **parasite** has a complex life cycle as shown in Figure 4.28. In 1955, the **World Health Organisation (WHO)** spent hundreds of millions of dollars in an attempt to eradicate malaria. The disease was attacked on several fronts:

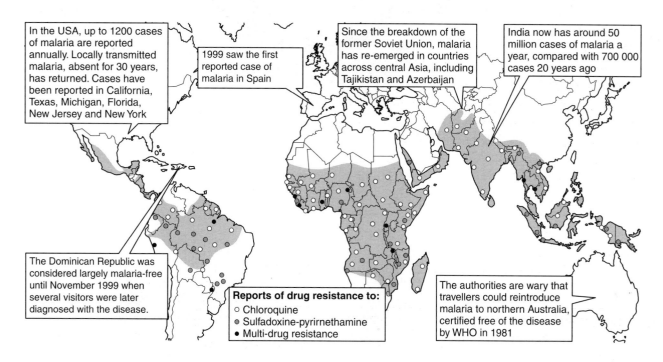

Figure 4.27 Global distribution of malaria.

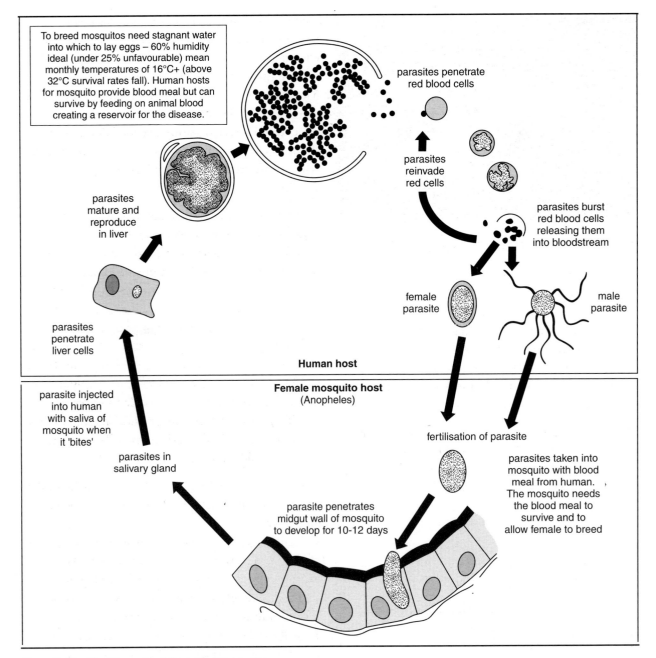

To breed mosquitos need stagnant water into which to lay eggs – 60% humidity ideal (under 25% unfavourable) mean monthly temperatures of 16°C+ (above 32°C survival rates fall). Human hosts for mosquito provide blood meal but can survive by feeding on animal blood creating a reservoir for the disease.

parasites penetrate red blood cells

parasites reinvade red cells

parasites burst red blood cells releasing them into bloodstream

parasites mature and reproduce in liver

female parasite

male parasite

parasites penetrate liver cells

Human host

Female mosquito host
(Anopheles)

parasite injected into human with saliva of mosquito when it 'bites'

parasites in salivary gland

fertilisation of parasite

parasites taken into mosquito with blood meal from human. The mosquito needs the blood meal to survive and to allow female to breed

parasite penetrates midgut wall of mosquito to develop for 10-12 days

Figure 4.28   The life cycle of the malarial parasite.

- First by the elimination of breeding areas. This was done by filling in and draining areas of stagnant water, spraying the areas with insecticides such as DDT and Malathion, spreading oil on the water surface, flushing streams to wash away the larvae and clearing trees and shrubs (which provide shelter for the mosquitos) from streams and ponds.

- Second by preventing the development of the larvae. This was done by introducing fish which eat the larvae, suffocating the larvae by spraying egg white on the breeding areas and using methods such as putting mustard seeds into paddy fields to drown the larvae by dragging them below the surface.

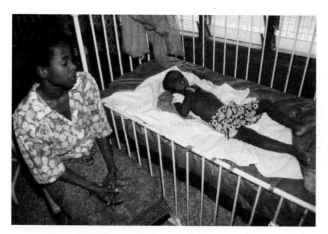

Figure 4.29 Child suffering from malaria.

- Third by treating sufferers with drugs such as quinine to kill the parasites within their bloodstream, killing mosquitos by spraying the interior of houses with insecticides such as DDT and running health education campaigns emphasising the ways in which the disease is transmitted and the need to seek early medical help especially for children.

The effectiveness of these methods varied. The disease was eradicated from Europe, and the former Soviet Union, most of the Middle East, North America and large areas of the Far East. Africa was virtually unaffected by the eradication methods and there was a resurgence of the disease in India in 1970. By 1969, the WHO realised that they could only exert some control over malaria, not eradicate it and abandoned their aim of worldwide eradication.

There were many problems which led to this situation. It proved extremely difficult to eliminate breeding areas. Mosquitos need only a small area of water in which to breed, for example, in water which has collected in an old tin can or even in an animal hoof print. Keeping an area clear of rubbish is not easy in urban slum areas nor is the elimination of animal tracks in rural areas. The use of drug treatments is another problem area as the parasites are becoming **resistant** to the present generation of antimalarial drugs, as a result of often the inappropriate use of these drugs in the past. In addition many countries are unable to finance a sustained antimalarial drug programme.

Despite the growing resistance to drugs it is still worthwhile treating the most vulnerable groups in endemic areas such as pregnant women and people visiting from areas where malaria is not found. Mosquitos are also becoming resistant to insecticides, and the use of the most common and effective insecticide DDT has greatly declined due to fears about its effects on the environment and the way in which it affects the long term health of people and wildlife.

Despite these difficulties, doctors believed that by 1980, malaria had been eliminated from many areas. In recent years however there has been a resurgence of malaria worldwide. This is due to a combination of factors. A rise in average temperatures worldwide has allowed malarial carrying mosquitos to extend their range beyond the tropical and subtropical areas to which they had been largely confined. Global climate change has also had its effects. For example in Kenya in 1998 there were four months of unusually heavy rain which created large areas of flood water providing an ideal breeding ground. The result was a 500% rise in cases of malaria in the affected areas with five children dying each day (Figure 4.29).

Another factor has been the collapse of public health programmes. For example, the break up of the USSR meant that the centrally controlled, large scale, antimalarial programmes of the past are no longer in place and malaria has reappeared in central Asia. Other health programmes in many developing countries have collapsed because debt repayments have meant that these countries are unable to maintain viable health services or afford effective treatments. In other areas malaria has been spread because of a breakdown in civil government due to war, natural disasters and famine. These all lead to large scale population movements. If these people are infected and move into an area free from malaria they will carry the disease into an area where people's immunity is non existent so that the disease can spread very rapidly with devastating effects on the victims.

The best advice at present is to try as far as possible to avoid being bitten. People should be especially careful at dawn and dusk when mosquitos are most active.

Precautions include having screens on doors and windows to prevent mosquitos entering the house, using insecticides or fumigants to kill any which have penetrated into the house and sleeping under a net treated with insecticide. Drug treatment is still advised although it is losing its effectiveness.

In the long term WHO would like to see the development of effective antimalarial drugs and in particular a **vaccine**. This is a difficult task as the malarial parasite is complex and is difficult to treat. The vaccine would have to be cheap, safe, easy to administer, be able to be used on infants and give life long immunity. It is extremely difficult to interest the private sector in such an expensive, technically difficult project with the prospect of poor returns on the finished product. Attempts are being made to speed up the process, for example Bill Gates has donated $50 million for research. Even if the attempt succeeds it is estimated that it will take at least 10–15 years for an effective vaccine to be developed. In the short term the prospects for malarial control are not good. The absence of the disease is not the same as the absence of risk and it is a worldwide issue not confined to a few areas.

Schistosomiasis (Bilharzia)

It is estimated that this disease affects 200 million people worldwide (Figure 4.30) causing 20 000 deaths a year mainly through kidney failure and bladder cancer. When the disease is untreated and becomes chronic it causes severe damage to the liver and the spleen. In areas with high incidence of the disease children are often particularly at risk. It is found in over 70 countries, in tropical and sub tropical areas with almost half of these in Africa alone. It is one of the major communicable diseases of the developing world, probably the second most prevalent after malaria.

The disease is caused by a parasitic infection associated with a flatworm or blood fluke. The parasite has a complicated life cycle as shown in Figure 4.31. It spreads when people come in contact with infected water while swimming, fishing, working on irrigated crops and whilst using the water for domestic or personal hygiene. People working in padi fields are particularly susceptible to schistosomiasis (Figure 3.32).

Figure 4.30   Global distribution of schistosomiasis.

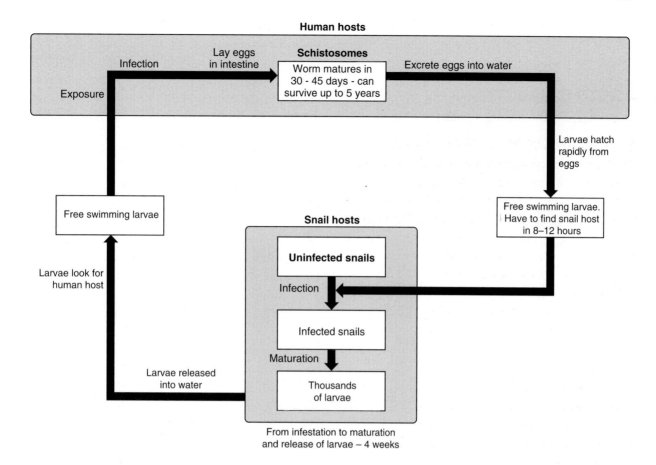

Figure 4.31   The life cycle of the schistosomiasis parasite.

The situation is made worse by poverty, inadequate or lack of public health facilities, rapid urbanisation with the creation of slum conditions, and population movements such as those associated with refugees fleeing conflicts and other disasters.

Figure 4.32   Workers in a padi field are at risk from Schistosomiasis.

A further reason for the continuing spread of the disease has been the creation of large freshwater reservoirs in high risk areas such as the Aswan Dam in Egypt (Figure 4.33), the Akosombo Dam in Ghana, the Kainji Dam in Nigeria and the Kariba Dam in Zimbabwe. Associated with these large scale projects there is usually an increase in irrigated land creating large areas of slowly moving water which allow the parasites to spread outwards into previously hostile environments.

As the disease causes **chronic** ill health it has a high economic impact as it debilitates workers and affects childrens' growth and school performance. Its impact is, however, often unrecognised both by the public and by health authorities. Affected people do not see it as a serious health problem as its symptoms are often unspecific and take a considerable time to develop. This often means that the disease goes untreated in its early stages so that by the time treatment starts the symptoms can be very serious.

Figure 4.33   Aswan Dam in Egypt.

Schistosomiasis can be attacked on several fronts. People who are infected can be treated with drugs. These used to be dangerous with numerous side effects. In recent years however new, safe and relatively cheap drugs have been developed. The problem for the developing countries is that, due to the large numbers of people infected, they cannot afford sufficient quantities of the drug and find distribution difficult. The host snails can be killed by using a molluscicide (chemical to kill snails) or by modifying the environmental conditions in which they breed, e.g. lining irrigation ditches with concrete and trapping the snails in filters or wire mesh. The introduction of adequate sanitation systems and other health care measures, including health education warnings about the dangers of being in contact with infected water, would also make inroads into the incidence of the disease.

Despite the fact that diagnosis and control of schistosomiasis is relatively straightforward, interest in control measures has decreased over the years so that the disease is once again spreading, even in areas such as Brazil and China where effective measures in the past had almost eliminated the occurrence of the disease. At present here is no vaccine against the parasite, so the priority is to educate people about the dangers to which they, especially children, are exposed in high risk areas.

Cholera

Cholera is believed to have originated in the Indian sub continent and spread outwards as world trade routes developed. It reached Europe in the nineteenth century causing many deaths.

By the end of the nineteenth century a better understanding of the causes of the disease and improved water and sanitation systems, meant that it had once again retreated to the Indian sub continent and south east Asia. In 1961 a new strain of the cholera bacteria emerged in south east Asia (the El Tor strain) and spread outwards from there probably carried by Muslim pilgrims to Mecca. At that stage the spread was relatively slow. In 1991 however there was a resurgence of cholera in South America, the first **epidemic** for almost a century, which between 1991 and 1995 infected 1 million people and killed 11 000.

The cholera bacteria is spread via water contaminated by human waste containing the bacteria. It has long been associated with unsanitary conditions i.e. a lack of adequate supplies of clean water and effective sewage systems in overcrowded urban areas and in poor rural areas. Such conditions are always made worse by natural disasters such as flooding or any circumstances which force large numbers of people into overcrowded unhygienic conditions such as refugee camps. Many people in these circumstances are forced to use contaminated water supplies for cooking, drinking and washing food. Cholera also has traditional links with the sea. The cholera organism thrives best in moderately salty water such as coastal estuaries and it has been found that it can tolerate open ocean conditions by surviving in some species of plankton. It is only found in freshwater where nutrient levels from organic pollution are high. Throughout history epidemics have started in coastal cities with unhygienic conditions.

Part of the blame for the recent resurgence in cholera is due to deteriorating water and sanitation systems in areas where governments do not have sufficient money to maintain them. Many developing world cities have poor living conditions and inadequate housing due to rapid population growth. Many people suffer from malnutrition and there is the added problem of street vendors who do not use refrigeration or clean water. In the countryside there is an increasing use of waste water for irrigation. None of this, however, is sufficient to explain the worldwide re-emergence of cholera.

Several theories have been advanced to explain the latest epidemics. Was the bacteria carried round the world in the ballast and sewage of cargo ships and discharged into coastal waters? Was the bacteria simply dormant in coastal waters waiting for favourable conditions? Was there a dormant reservoir in the oceans waiting until the conditions for expansive growth were present? The discovery that the cholera organism could survive in some species of plankton does suggest a possible route by which cholera could spread rapidly over a wide geographical area carried by ocean currents. Changing sea conditions such as the warming of surface waters by a change in the course of the El Niño current could allow an explosion of plankton numbers (called a **plankton bloom**) in coastal waters, and this combined with nutrients from sewage could re-awaken the cholera bacteria.

Cholera is treatable by drugs and oral rehydration therapy (replacing liquid and salts lost through sickness and diarrhoea). If it remains untreated, death is common due to acute dehydration, with children particularly at risk. The most effective way to control the disease is to provide safe, clean, piped and chlorinated water, and where that is not possible to dig deeper wells and introduce people to simple water purification techniques. The provision of effective sanitation systems is also vital to prevent contamination of water supplies. The root causes of overcrowding and poverty also have to be tackled in order for effective control measures to be put into place and maintained over the long term.

In the past people who had recovered from an attack of cholera were considered to be immune. However the bacteria has since mutated and these people are now also at risk from re-infection. It makes the development of an effective vaccine unlikely, so that control measures become even more important. The fact that plankton blooms may be implicated in the spread of cholera is also of importance. Plankton blooms are increasing in frequency as global temperatures rise, nutrient levels increase due to increasing populations and atmospheric carbon dioxide levels increase. It could be possible to set up early warning systems so that people in coastal areas could take extra precautions, such as the chlorination of water supplies, if a plankton bloom approached.

It is unlikely however that much progress will be made in dealing with cholera until there is an improvement in the supply of water and sanitation worldwide.

The provision of primary health care

Throughout the developing world, governments are unable to afford the elaborate health care systems found in the wealthier parts of the world. In an effort to improve the health of their citizens they have instead adopted a number of strategies classified as **primary health care**.

One of the earliest attempts to provide basic health care to large numbers of rural dwellers was set up in China in the 1960s. The communist government was faced with a countryside which had virtually no access to health care. Most villages lacked medical facilities and the nearest health centres could be up to three days walk away. Even if they could be reached, patients would find that they were poorly equipped and run by inadequately trained staff.

To deal with this situation large numbers of local people were trained as medical auxiliaries called **barefoot doctors**. They were given basic medical training which allowed them to treat straightforward illnesses and to refer more serious cases on to better equipped hospitals. The typical medical auxilary was a part time health worker. The rest of the time they supported themselves in jobs such as farming. The costs of running the medical service was met by the community and a combination of Western and traditional Chinese medical practices were used.

In the early years the system was very successful at making inroads into the poor state of health in the countryside. In recent years however the service has faced considerable pressures. Despite the numbers trained it proved impossible to provide enough health care workers to satisfy the needs of rural communities. In addition, many were unable to continue paying for the training of auxiliaries and the upkeep of the health centres, so the system began to break down. To try to deal with this, the Chinese government started to encourage **medical auxiliaries**, traditional doctors and hospital physicians to set up in private practice.

However, the introduction of private practice has not been to the benefit of large numbers of people. The countryside is generally too poor to attract health workers who prefer to settle in the wealthier suburban areas. Once again many rural communities are suffering from a lack of effective, affordable health care. Unless the Chinese governmnet can find the solution to these problems the situation is likely to deteriorate.

Despite this failure, the pattern has been adopted by many countries in the developing world as they struggle to improve the health of those living in rural areas. Many of these schemes have also run into difficulties. Burdened with debt repayments and forced to make structural readjustments by the World Bank, many countries were unable to fund their health services. In Tanzania, for example, medical treatment is no longer free and the ordinary citizen cannot afford to pay for the treatments. In other areas such as Mozambique, civil war and natural disasters destroyed what had been a highly effective health provision. Many schemes, however, still exist often funded by charities such as Christian Aid.

The charities tend to work with local organisations which are aware of local needs. In India for example Christian Aid works with local groups to train health volunteers and equip them with basic medical kits which are simple and cheap to provide and re-equip. Because the health volunteers are local people they are known and trusted by the communities which they serve. An integral part of this partnership is an emphasis on health education. The method is still an extremely cost effective methods of reaching large numbers of people in the countryside allowing them to build a better future for themselves and their children.

The other main emphasis is to concentrate on low cost, low technological solutions which encourage self help and cut reliance on the often expensive foreign medicines. Attempts are been made to identify medicines which can be produced locally often based on traditional practices which have proved effective in the past. An example is the use of Oral Rehydration Therapy (ORT – a solution of water, sugar and salt) to deal with the dehydration resulting from acute diarrhoea – one of the major causes of infant mortality in the developing world (Figure 4.34).

Figure 4.34   Oral Rehydration Salts for the treatment of diarrhoea.

The establishment of clinics to provide advice on nutrition, family planning and health education although more expensive will bring benefits in the long term as will large scale child immunisation programmes which usually need foreign funding in order to be effective. All these measures including foreign intervention are needed if inroads are to be made into the poor levels of health provision throught the developing world.

Case Study: Chogoria Hospital – Kenya

Chogoria Hospital is located 190 km north of Nairobi on the eastern slopes of Mount Kenya (Figures 4.35 & 4.36) at an altitude of 1540 m. The area around Mount Kenya at 2000 m is well watered (2000 mm of rainfall per annum) and has rich volcanic soils. Further east near the Tana River the land, at a height of 300 m, is flat and arid. In the well-watered areas farmers grow maize, millet, potatoes, fruit and vegetables for their own use and tea, coffee and pineapples as cash crops for export. In the driest areas people rely on subsistence farming and pastoralism and are generally much poorer with greater health problems. There are few tarmaced roads, most are dirt roads which are difficult to pass in the wet season (Figure 4.37). The hospital serves approximately 555 000 people with a population increase of over 2% per annum.

The hospital was established in 1922 as a Church of Scotland mission hospital and in 1923 treated 7000 outpatients (Figure 4.38). The hospital grew steadily through the 1920s and 1930s until by 1936 it treated 760 inpatients and 11 606 outpatients. Numbers fell during the Second World War as it was affected by shortages of personnel and equipment.

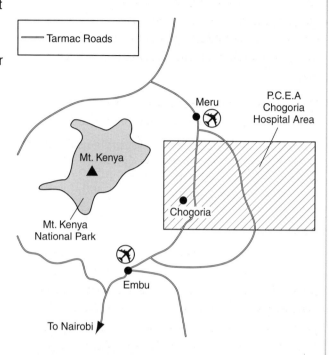

Figure 4.36 Location of Chogoria Hospital.

Expansion continued after the war with a greater range of services on offer. The Mau Mau uprising affected the work of the hospital but work continued to expand, and once the emergency was over, there were 312 beds available for inpatients as well as a range of medical services.

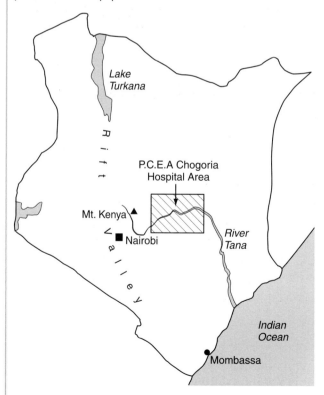

Figure 4.35 Location of Chogoria Hospital.

Figure 4.37 In the rainy season, patients may find it difficult to reach the hospital.

Figure 4.38   The clinic post where Chogoria Hospital started in 1924.

These included: surgical ward, maternity ward, TB treatment, dental and eye clinics. The hospital was also the centre for training nurses and midwives (Figures 4.39–4.41).

Figure 4.39   Surgical ward in Chogoria Hospital.

Figure 4.40   Mural on the wall of the community health building at Chogoria Hospital.

Figure 4.41   Paediatric ward in Chogoria Hospital.

After Kenya gained independence the hospital continued its work in conjunction with the Kenyan government. By this time the hospital was beginning to experience considerable financial problems.

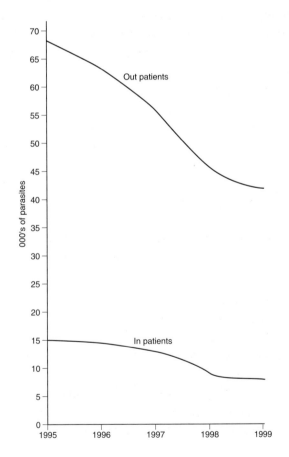

Figure 4.42   Number of patients being treated in Chogoria Hospital.

Since the start of the hospital's work, there had always been a policy of charging a small fee, although no one in need and unable to pay had ever been turned away. The problems became more severe due to several factors. The Kenyan government had promised free health care which in the event it was unable to deliver, apart from to outpatients and children. The financial problems became acute in the 1980s and 1990s as the country was suffering from the effects of debt repayment and the falling world prices for its major exports. In 1993 inflation was 100% and the hospital was forced to increase fees by 50%. This has led to a steady decline in the numbers of patients treated (see Figure 4.42) and bed occupancy which had been as high as 160% in 1970 had fallen to 65% in 1999.

The need for outreach into the surrounding rural areas has been seen from the beginning and by 1970 there were eight dispensaries staffed by hospital trained nurses running day clinics. This network has gradually expanded until today there are now 32 clinics run by area health committees and offering a range of services. The clinics treat those who come in with diseases such as TB and malaria. They also undertake preventative work, for example, via a child immunisation programme. Ante and post natal advice is on offer as well as midwifery services. There is also a range of health education programmes on offer including family planning advice and information on how to prevent the spread of HIV and AIDS. Once again attendance at these clinics is on the decline (Figure 4.43).

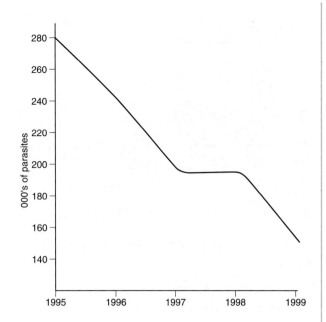

Figure 4.43   The decline in the number of patients treated in rural clinics.

The major illness is still malaria, however is is becoming resistant to the less expensive drugs used in the past, leaving only the more expensive drugs as possible treatments. There has been a worrying increase in the cases of TB and the problem of HIV and AIDS is on the increase. The greatest concern is the fact that an increasing proportion of the population can no longer afford hospital treatment and are either going without treatment or turning to unqualified healers.

Questions

9   Describe the physical and human factors which affect the incidence of diseases throughout the world.

10   For each of the following diseases; malaria, schistosomiasis, cholera:
   a Describe the factors which affect its incidence and transmission.
   b Describe the effects on the economies of the areas involved.
   c Describe the measures taken to control and/or eradicate the disease.
   d Comment on the effectiveness of these methods.

11   Explain what is meant by primary health care and explain why it has been adopted as a health care policy by many developing countries.

12   Give examples of primary health care and comment on the appropriateness of such schemes in the context of the developing world.

13   Referring to specific examples from the text, identify the strengths and the problems of some primary health care policies.

14   Outline the work of charities in proving primary health care in many developing countries.

 **Key terms and concepts**

Summary

Having worked through this section, you should now know:

- there are marked inequalities in social and economic development globally, regionally and nationally

- levels of development can be measured by a range of indicators – economic, social and composite

- within the developing world, there are considerable differences in levels of development both between and within countries

- the incidence of disease throughout the world is due to a combination of physical and human factors which include climate, water supply, levels of wealth and nutrition

- malaria, schistosomiasis (bilharzia) and cholera each have distinctive features leading to their methods of transmission and distribution

- there are a variety of strategies in place to control disease and its impact. These have had varying degrees of success.

The Examination (Paper 2 Applications of Geography)

The Examination (Paper 2 Applications of Geography)

The structure of the examination

For the final examination in Higher Geography there are two papers.

- **Paper 1** covers the eight topics from Core Geography
- **Paper 2** covers the Applications.

The structure of Paper 2

In all there are six applications, four of which have been covered in this textbook. The paper will have one question for each application and will be divided into two sections.

- **Section 1:** Rural Land Resources; Rural Land Degradation; River Basin Management★
- **Section 2:** Urban Change and its Management; European Regional Inequalities★ Development and Health.

(★Denotes Applications not covered in this textbook)

The examination lasts for 1 hour 15 minutes, and you will be required to answer **two** questions: **one** from **Section 1**, and **one** from **Section 2.**

Questions will be divided into a number of parts but most parts will generally carry more marks than those in Paper 1 (Core Geography) and will therefore require longer and more detailed answers.

Choosing your questions

During your Higher Geography course you will have studied at least three of the six Applications, so that in Paper 2 of the examination you will have some choice, but choose carefully!

- Ensure that you are clear about which Applications you have studied.

- Quickly read through the questions for all three of the Applications which have been included in your course and make your choice of the two that you will feel most confident about answering.
- Most questions begin with a shorter introductory 'starter' which carries approximately 5 marks, **but** look in particular at those parts of the questions which carry higher mark allocations – you will need to answer these in considerable depth, using case study material and specific examples.
- It is very tempting to go for those questions which include familiar terms like 'primary health care' or 'National Park' but read the **entire** question before you start, to make sure that you can really answer it all.

Answering the questions

The examiners are not trying to catch you out, but rather will award marks when you give **specific** and **direct** answers to the questions which are set.

- Ensure that you read the question properly and that you tailor your answer accordingly. As well as developing the main theme of the question in your answer, make sure that you have underlined the **key instructions.** These are usually:

account for – explain the cause of
analyse – look at the issues, break them down and show how they interrelate
annotate – add labels to show key features
assess – weigh up the importance of
comment – write an explanation
compare – point out similarities and differences
contrast – emphasise the differences only
describe – state the main characteristics or features
discuss – present arguments for and against an issue
evaluate – review whether something has been successful

explain – give reasons for

illustrate – give an example which could be a case study, sketch map or diagram

outline – give the main features

suggest – put forward a reason or an idea

- Although time is short in the examination, a very brief outline plan of your answer should take no more than a few minutes and will help structure your answer.

- Make sure that you answer all parts of the question. Some may require a number of separate approaches within a single answer:

e.g. *'social, economic and environmental factors'*

- Make good use of the case study material which has been included as part of your course. In this paper, your ability to use relevant case studies is very important. Answers which are entirely abstract and fail to utilise real world material will not score highly.

- Use any resource material which has been provided as part of the question. Although at first sight some graphs or diagrams may look complex, they are often very useful in providing you with broad headings to include in your answer.

Examination style questions

Rural Land Resources

1 a Study Reference Diagram Q1 which shows part of the South Downs area of southern England.

Describe and explain the structure, relief and drainage of both the upland and lowland areas shown on the diagram. **10**

b The South Downs have been designated as an Environmentally Sensitive Area (ESA). Explain the purposes of ESAs and how they benefit both conservation and farming interests. **5**

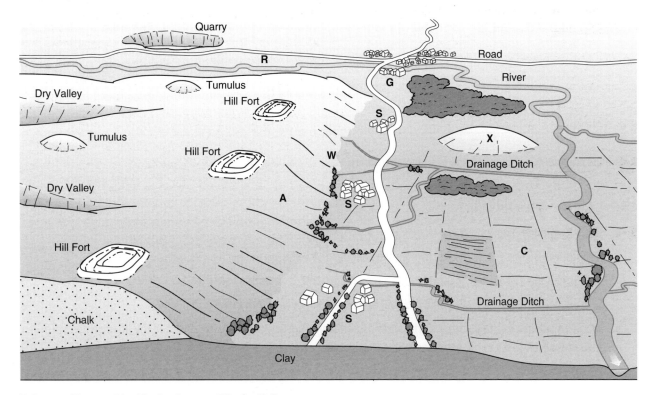

Reference Diagram Q1   The landscape of the South Downs.

c Land use conflicts and pressures on areas such as the South Downs continue to increase. For the South Downs, or any other upland area of the UK which you have studied:

    i identify the main conflicts which have taken place

    ii suggest ways in which such conflicts can be resolved.   10

    (25)

2 a Study Reference Diagram Q2 which shows part of the glaciated landscape of the Cairngorm Mountains.

    With the aid of **annotated** diagrams explain how the main features of such a landscape were formed.   10

  b With reference to the Cairngorms, or any other UK upland area which you have studied:

    i explain the social and economic opportunities created by the landscape

    ii give examples of environmental conflicts which may arise in the chosen area.   10

  c Outline the arguments for and against the creation of National Parks in Scotland.   5

    (25)

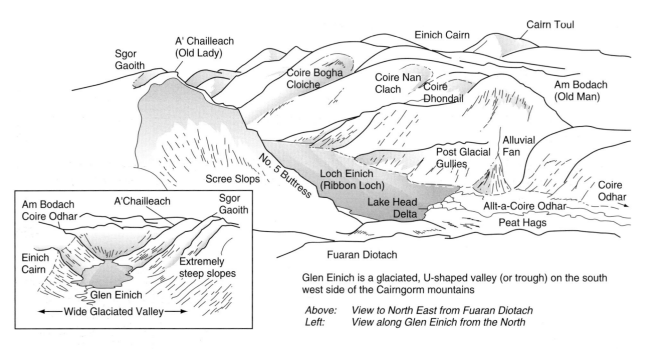

Glen Einich is a glaciated, U-shaped valley (or trough) on the south west side of the Cairngorm mountains

Above: *View to North East from Fuaran Diotach*
Left: *View along Glen Einich from the North*

Reference Diagram Q2   The landscape of a glaciated valley.

Rural Land Degradation

3 a Study Reference Diagram Q3

    For **either** the Great Plains or the savanna grasslands, and referring to specific examples:

    i describe the human activities which have resulted in land degradation   8

    ii explain in detail the effects of the removal of vegetation cover on both land and people.   8

  b Choosing specific examples from North America and Africa north of the Equator, describe the effectiveness of measures taken to try to conserve the soil in these areas.   9

    (25)

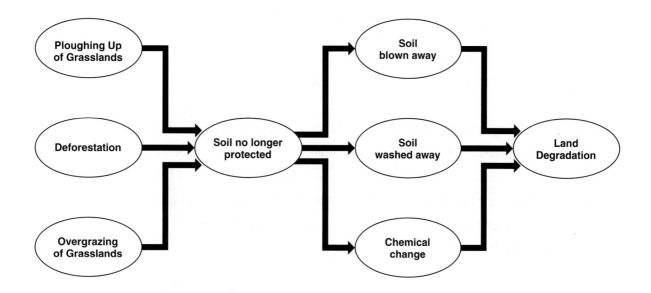

Reference Diagram Q3   The processes involved in land degradation.

4 a Study Reference Map Q4a (page 180).

Describe the annual rainfall patterns and show how variations in rainfall may lead to land degradation.   5

b For either the Great Plains of USA or the African Sahel describe the human factors which have led to rural land degradation.   10

c Reference Diagram Q4b illustrates the work of the Tennessee Valley Authority in the USA. Choose any four methods of soil conservation shown in the diagram and explain how each helps to conserve soil and reduce land degradation.   10

(25)

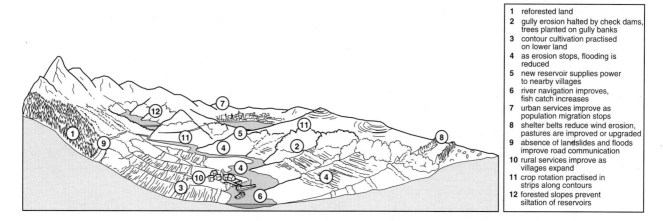

1 reforested land
2 gully erosion halted by check dams, trees planted on gully banks
3 contour cultivation practised on lower land
4 as erosion stops, flooding is reduced
5 new reservoir supplies power to nearby villages
6 river navigation improves, fish catch increases
7 urban services improve as population migration stops
8 shelter belts reduce wind erosion, pastures are improved or upgraded
9 absence of landslides and floods improve road communication
10 rural services improve as villages expand
11 crop rotation practised in strips along contours
12 forested slopes prevent siltation of reservoirs

Reference Diagram Q4b   The work of the Tennessee Valley Authority.

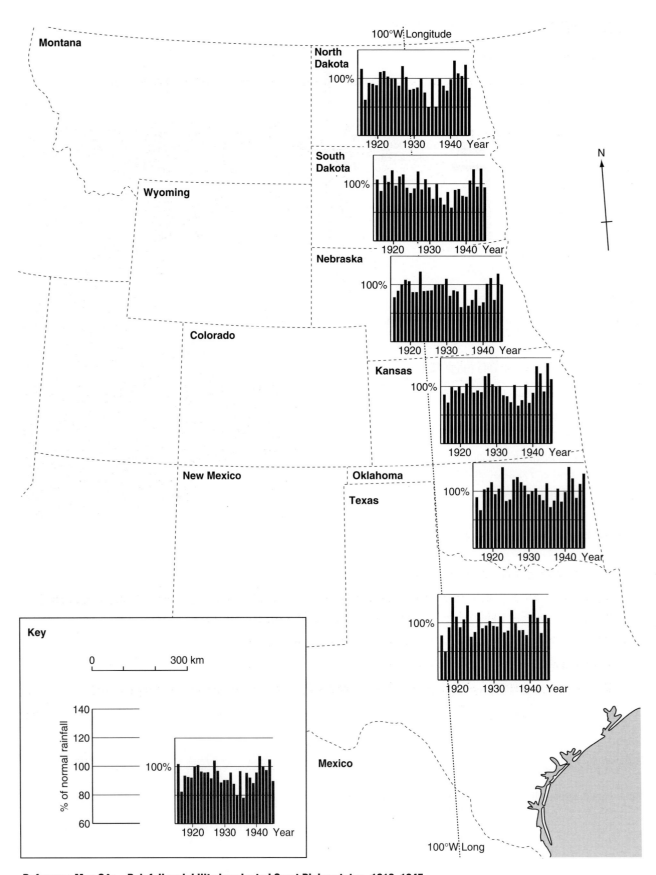

Reference Map Q4a   Rainfall variability in selected Great Plains states, 1916–1945.

Urban Change and its Management

5 a Study Reference Map Q5a.

Describe and explain the distribution of major cities in Canada, or in another **developed** country you have studied.

5

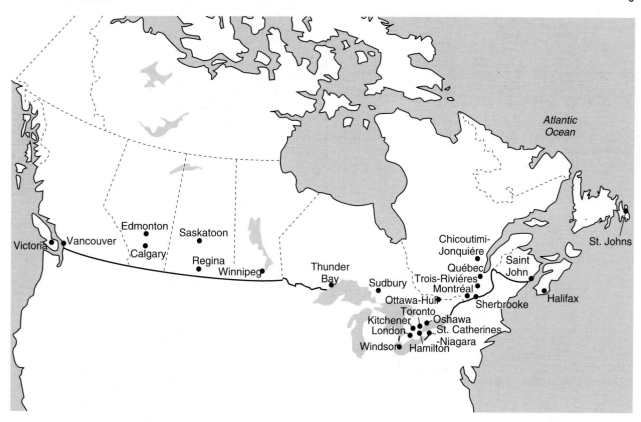

Reference Diagram Q5a   Canadian cities.

b Study Reference Diagram Q5b (page 182).

Apply the models of processes influencing the changing location of urban land uses to a city from the **developed world** which you have studied, and comment on any similarities and differences.

10

c With reference to a city which you have studied in the **developing world**:

i describe and account for its rapid growth

ii identify the social, economic and environmental problems which have resulted from its rapid growth.

10

(25)

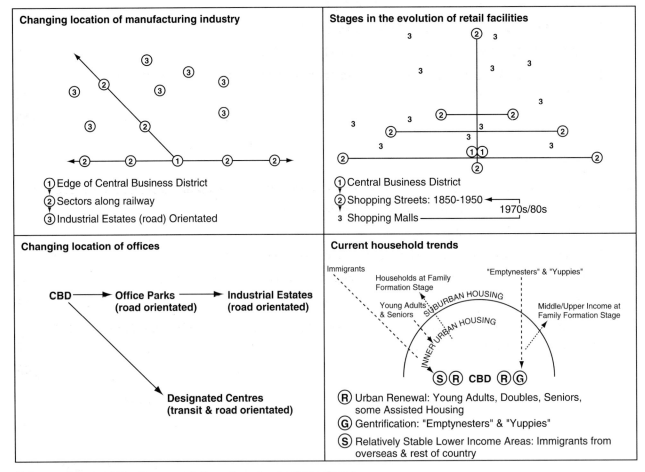

Changing location of manufacturing industry

①Edge of Central Business District
②Sectors along railway
③Industrial Estates (road) Orientated

Stages in the evolution of retail facilities

①Central Business District
②Shopping Streets: 1850-1950
3 Shopping Malls ──── 1970s/80s

Changing location of offices

CBD ──► Office Parks (road orientated) ──► Industrial Estates (road orientated)

CBD ──► Designated Centres (transit & road orientated)

Current household trends

Immigrants
Households at Family Formation Stage
Young Adults & Seniors
"Emptynesters" & "Yuppies"
Middle/Upper Income at Family Formation Stage
SUBURBAN HOUSING
INNER URBAN HOUSING
ⓈⓇ CBD ⓇⒼ

Ⓡ Urban Renewal: Young Adults, Doubles, Seniors, some Assisted Housing
Ⓖ Gentrification: "Emptynesters" & "Yuppies"
Ⓢ Relatively Stable Lower Income Areas: Immigrants from overseas & rest of country

Reference Diagram Q5b   Processes influencing urban land use change.

6 a Study Reference Diagram Q6.

Compare and contrast the land use pattern of a city which you have studied from the **developing world** with the model, highlighting similarities and differences.

5

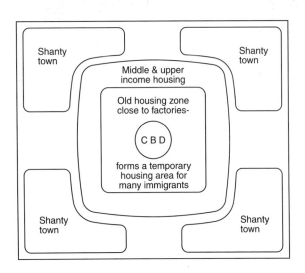

Shanty town

Shanty town

Middle & upper income housing

Old housing zone close to factories-

C B D

forms a temporary housing area for many immigrants

Shanty town

Shanty town

Reference Diagram Q6   Model of a city in the developing world.

b With reference to any named city from the **developing world**:

   i describe the main problems associated with shanty towns

   ii examine the solutions proposed, and comment on their effectiveness. 10

c Referring in detail to a named city in the **developed world**, describe and explain the changes which have taken place in housing, industry, shopping and transport since 1950. 10

(25)

Development and Health

7 a Describe and explain the differences which may exist in levels of development **within** countries in the **developing world**. Your answer should refer to at least one specific country. 5

b Study Reference Maps Q7a and Q7b.

   Choosing either malaria or another named disease in the **developing world**:

   i outline the physical and human factors which contribute to the spread of the disease

   ii describe the methods used to control the disease and comment on their success. 12

c Using specific examples of primary health care, explain why it is an appropriate policy for improving health care provision in many **developing countries**. 8

(25)

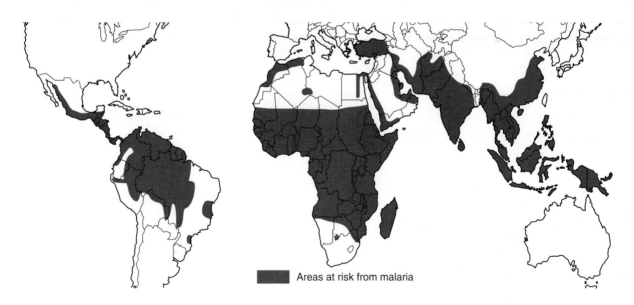

Areas at risk from malaria

Reference Diagram Q7a   Areas at risk from malaria.

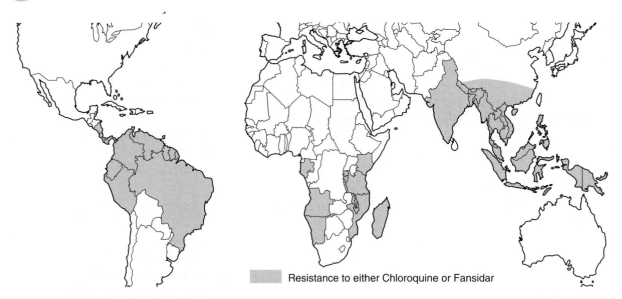

Resistance to either Chloroquine or Fansidar

Reference Diagram Q7b   Resistance to selected anti malarial drugs.

8 a Give **three** examples of indicators which might be used to illustrate a country's level of development, and for each comment on their value for this purpose.   6

   i Explain why such indicators may not provide an accurate picture of the true standards of living **within** a country.

   ii Give reasons for the differences in levels of development which exist **between** different countries in the **developing world**.   7

  b Referring to specific examples, outline the importance of:

   i primary health care
   ii international organisations

  in attempting to combat the effects of any **one** named disease in the **developing world**.   12

(25)

INDEX

(entries in italics refer to illustrations;
entries in bold refer to tables)

Figure 1.1 Scottish Highlands.

Figure 1.2 East Anglia.

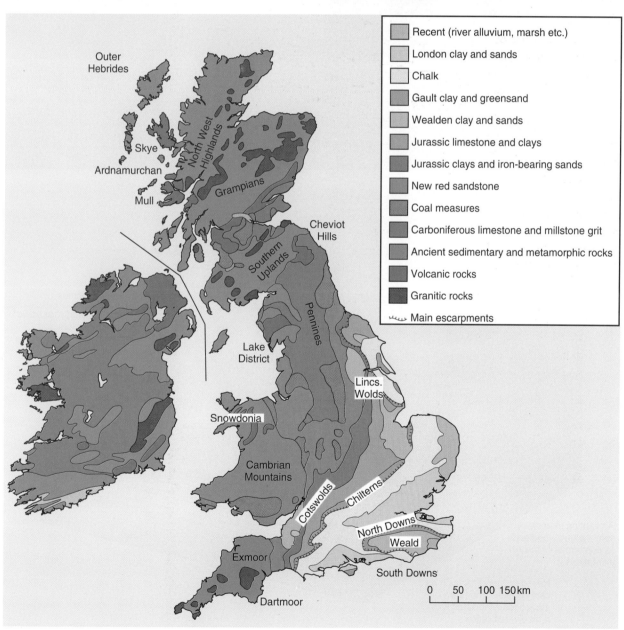

Figure 1.3 The geology of the British Isles.

Figure 1.8   General view of the Cotswolds.

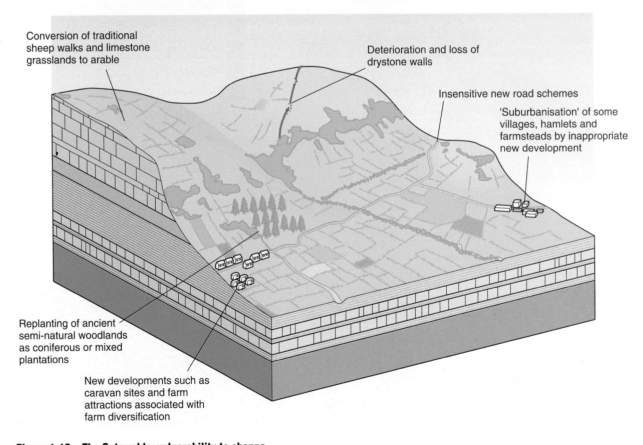

Conversion of traditional sheep walks and limestone grasslands to arable

Deterioration and loss of drystone walls

Insensitive new road schemes

'Suburbanisation' of some villages, hamlets and farmsteads by inappropriate new development

Replanting of ancient semi-natural woodlands as coniferous or mixed plantations

New developments such as caravan sites and farm attractions associated with farm diversification

Figure 1.12   The Cotswolds: vulnerability to change.

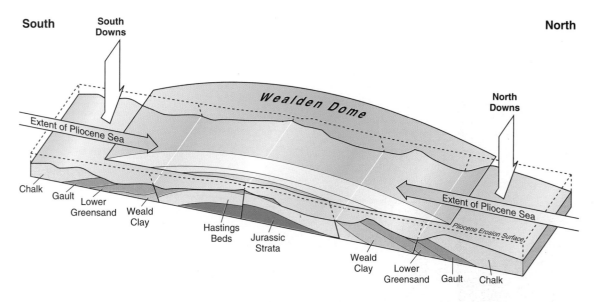

Figure 1.15   Cross-section of the Wealden Dome and escarpments.

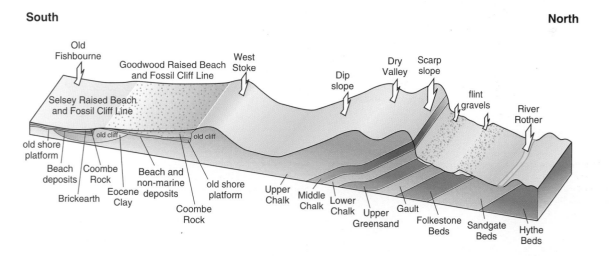

Figure 1.16   The physical structure of the chalklands: escarpments, dip slope and valleys.

Figure 1.18   Landscape of the South Downs.

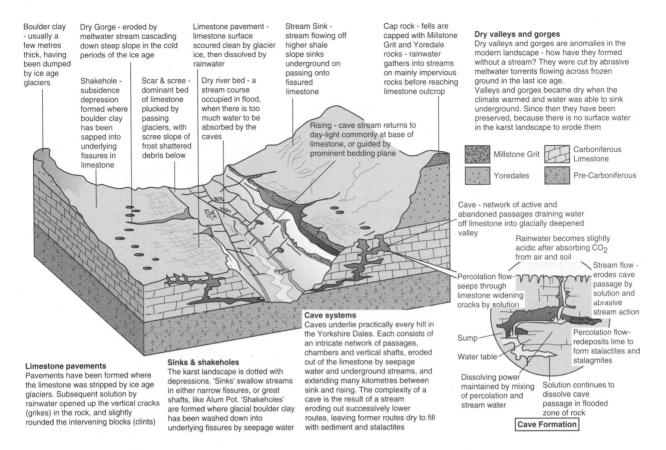

Boulder clay - usually a few metres thick, having been dumped by ice age glaciers

Dry Gorge - eroded by meltwater stream cascading down steep slope in the cold periods of the ice age

Limestone pavement - limestone surface scoured clean by glacier ice, then dissolved by rainwater

Stream Sink - stream flowing off higher shale slope sinks underground on passing onto fissured limestone

Cap rock - fells are capped with Millstone Grit and Yoredale rocks - rainwater gathers into streams on mainly impervious rocks before reaching limestone outcrop

Dry valleys and gorges
Dry valleys and gorges are anomalies in the modern landscape - how have they formed without a stream? They were cut by abrasive meltwater torrents flowing across frozen ground in the last ice age.
Valleys and gorges became dry when the climate warmed and water was able to sink underground. Since then they have been preserved, because there is no surface water in the karst landscape to erode them

Shakehole - subsidence depression formed where boulder clay has been sapped into underlying fissures in limestone

Scar & scree - dominant bed of limestone plucked by passing glaciers, with scree slope of frost shattered debris below

Dry river bed - a stream course occupied in flood, when there is too much water to be absorbed by the caves

Rising - cave stream returns to day-light commonly at base of limestone, or guided by prominent bedding plane

Millstone Grit

Yoredales

Carboniferous Limestone

Pre-Carboniferous

Cave - network of active and abandoned passages draining water off limestone into glacially deepened valley

Rainwater becomes slightly acidic after absorbing CO_2 from air and soil

Percolation flow- seeps through limestone widening cracks by solution

Stream flow - erodes cave passage by solution and abrasive stream action

Cave systems
Caves underlie practically every hill in the Yorkshire Dales. Each consists of an intricate network of passages, chambers and vertical shafts, eroded out of the limestone by seepage water and underground streams, and extending many kilometres between sink and rising. The complexity of a cave is the result of a stream eroding out successively lower routes, leaving former routes dry to fill with sediment and stalactites

Sump

Water table

Percolation flow- redeposits lime to form stalactites and stalagmites

Dissolving power maintained by mixing of percolation and stream water

Solution continues to dissolve cave passage in flooded zone of rock

Cave Formation

Limestone pavements
Pavements have been formed where the limestone was stripped by ice age glaciers. Subsequent solution by rainwater opened up the vertical cracks (grikes) in the rock, and slightly rounded the intervening blocks (clints)

Sinks & shakeholes
The karst landscape is dotted with depressions. 'Sinks' swallow streams in either narrow fissures, or great shafts, like Alum Pot. 'Shakeholes' are formed where glacial boulder clay has been washed down into underlying fissures by seepage water

Figure 1.29   Karst scenery.

Figure 1.33   Gordale scar.

Figure 1.34   Limestone pavement in the North Yorkshire National Park.

Figure 1.35   Traditional Yorkshire Dales farmland.

Figure 1.43   Swinden Quarry.

Figure 1.44   Swinden Quarry after restoration.

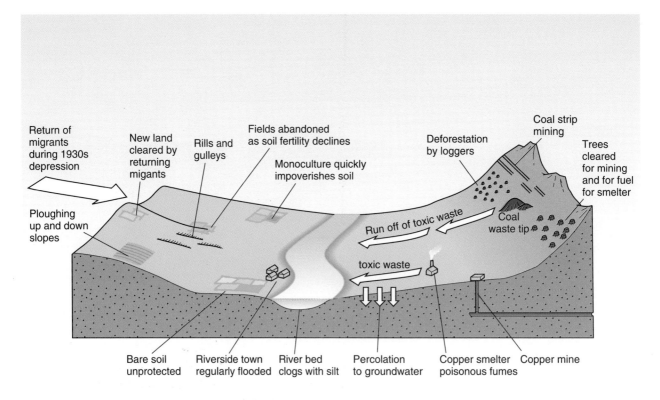

Figure 2.21   Causes of land degradation in the Tennessee Valley.

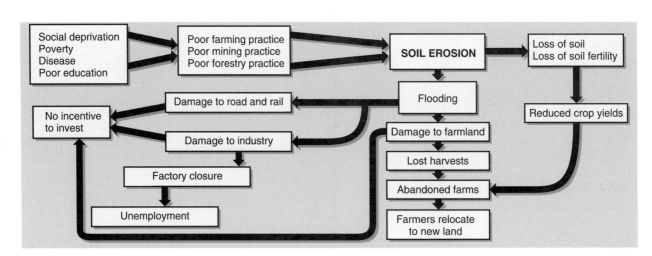

Figure 2.23   The causes and consequences of the Tennessee Valley Authority.

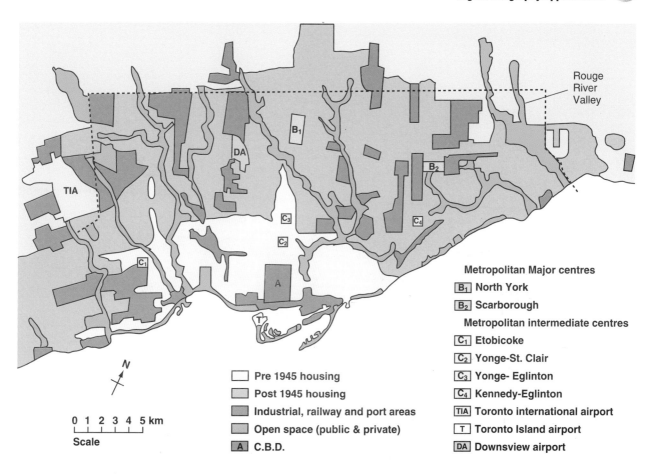

Figure 3.13   Land use in Toronto.

Metropolitan Major centres
- B₁ **North York**
- B₂ **Scarborough**

Metropolitan intermediate centres
- C₁ **Etobicoke**
- C₂ **Yonge-St. Clair**
- C₃ **Yonge- Eglinton**
- C₄ **Kennedy-Eglinton**
- TIA **Toronto international airport**
- T **Toronto Island airport**
- DA **Downsview airport**

- ☐ Pre 1945 housing
- ☐ Post 1945 housing
- ☐ Industrial, railway and port areas
- ☐ Open space (public & private)
- A C.B.D.

0 1 2 3 4 5 km
Scale

N

Figure 4.2   A rural landscape in the Developing World.

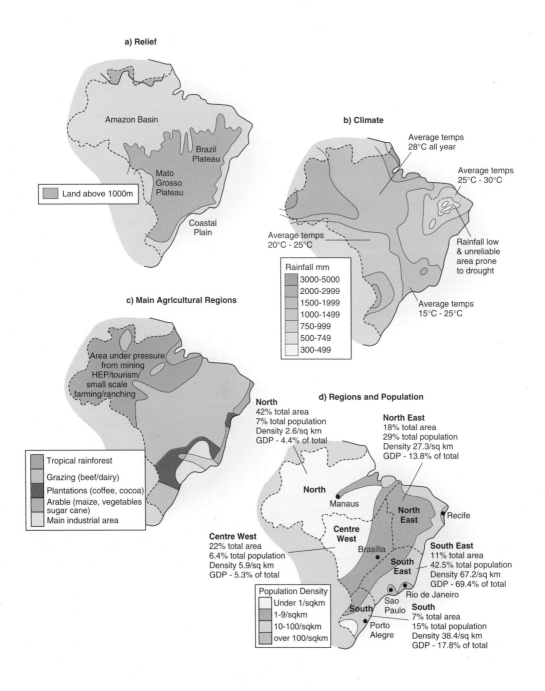

a) Relief

Amazon Basin

Brazil Plateau

Mato Grosso Plateau

Coastal Plain

Land above 1000m

b) Climate

Average temps 28°C all year

Average temps 25°C - 30°C

Average temps 20°C - 25°C

Rainfall low & unreliable area prone to drought

Average temps 15°C - 25°C

Rainfall mm
3000-5000
2000-2999
1500-1999
1000-1499
750-999
500-749
300-499

c) Main Agricultural Regions

Area under pressure from mining HEP/tourism/ small scale farming/ranching

Tropical rainforest
Grazing (beef/dairy)
Plantations (coffee, cocoa)
Arable (maize, vegetables sugar cane)
Main industrial area

d) Regions and Population

North
42% total area
7% total population
Density 2.6/sq km
GDP - 4.4% of total

North East
18% total area
29% total population
Density 27.3/sq km
GDP - 13.8% of total

Centre West
22% total area
6.4% total population
Density 5.9/sq km
GDP - 5.3% of total

South East
11% total area
42.5% total population
Density 67.2/sq km
GDP - 69.4% of total

South
7% total area
15% total population
Density 38.4/sq km
GDP - 17.8% of total

North

Manaus

North East

Recife

Centre West

Brasilia

South East

Rio de Janeiro

Sao Paulo

South

Porto Alegre

Population Density
Under 1/sqkm
1-9/sqkm
10-100/sqkm
over 100/sqkm

Figure 4.26